Making Workbenches

Planning · Building · Outfitting

Sam Allen

Sterling Publishing Co., Inc. New York

Acknowledgments

I would like to thank the following people and companies for their help with the preparation of this book: Virginia Allen, Paul Allen, John Allen, Record Tools, Inc., Adjustable Clamp Company, Veritas Tools, Inc., and Yost Manufacturing Company.

Library of Congress Cataloging-in-Publication Data

Allen, Sam.
    Making workbenches : planning, building, outfitting / Sam Allen.
      p.   cm.
    Includes index.
    ISBN 0-8069-0535-2
    1. Workbenches. 2. Furniture making. 3. Workshops—Equipment and supplies. I. Title.
TT197.5.W6A45 1995
684'.08—dc20
                                          95-4875
                                          CIP

10 9 8 7 6 5 4

Published by Sterling Publishing Company, Inc.
387 Park Avenue South, New York, N.Y. 10016
© 1995 by Sam Allen
Distributed in Canada by Sterling Publishing
% Canadian Manda Group, One Atlantic Avenue, Suite 105
Toronto, Ontario, Canada M6K 3E7
Distributed in Great Britain and Europe by Cassell PLC
Wellington House, 125 Strand, London WC2R 0BB, England
Distributed in Australia by Capricorn Link (Australia) Pty Ltd.
P.O. Box 6651, Baulkham Hills, Business Centre, NSW 2153, Australia
Manufactured in the United States of America

Sterling ISBN 0-8069-0535-2

Contents

Introduction

Early in my woodworking career, I worked in a large shop. There were six workbenches in the shop. Although they were massive and well-built, they had several deficiencies that frustrated me. Their tops were large, but there were no dog holes. They had only one vise mounted on their fronts, and the edges of their tops were rounded to a ¾-inch radius. That made it difficult to use a bench hook or a mitre box.

The large top was great for assembling large pieces of furniture, but it was difficult to work on smaller parts. To hold work on one of these benches, I usually used C-clamps or hand screws. The clamps would invariably get in the way as I was working, so I would have to move them several times. I would often attach cleats to the bench top to hold a particular part. The top was scarred with many nail holes from the temporary cleats.

I worked at those benches for five years. I gradually made improvements to them, but they still had fundamental flaws. When I set up my own shop, I wanted a better bench. I researched many traditional and modern designs and made my own. I liked this one better, but as I used it I still found ways to improve it. I have built several benches since then; each time I get closer to the ideal bench for me.

No one bench design is right for everyone. In this book I present basic information about workbenches, so that you can make an informed choice when you decide to either buy a commercial bench or build your own. In Chapter One, I describe the different types of workbench. In Chapters Three and Four, I present designs for two workbenches that you can build. Both are based on traditional designs. I have found that the basic problems of bench design have all been solved hundreds of years ago, and that the traditional designs are the best ones. Also included in the following pages are ways to use workbench accessories to get the most out of your bench, a description of the different types of vises, which many woodworkers consider an essential part of the workbench, as well as techniques for incorporating storage space in your bench and outfitting it with tools and accessories.

This book is the result of more than 25 years of research, trial and error, and experience. I hope that it will help you find the right bench for your needs as quickly and easily as possible.

Sam Allen

CHAPTER ONE

Workbench Fundamentals

A WORKBENCH IS MORE than a table; it is a tool. A well-designed and built workbench (Illus. 1-1) provides a stable working platform. Workbench accessories turn a bench into a sophisticated clamping tool. Vises, dogs, stops, and hold-downs make it possible to secure a board to the bench top. The ability to clamp the work to the bench makes the job easier and safer, because you have both hands free and there is less chance of the work slipping.

USING A WORKBENCH

Shop space always seems to be at a premium. To gain floor space, you usually move the workbench against the wall, but workbenches are designed to be free-standing. Try moving your workbench out into the middle of your shop and you will notice how much more versatile it is when you can walk around it and get at the work from all sides.

The workbench should be the focus of the shop, with the tools clustered around it. Even in a small shop, the work flow will be more efficient if you place the workbench in the middle (Illus. 1-2). Following are several woodworking techniques that can be done on a workbench.

Planing

Most of the features of the traditional bench designs were developed to facilitate hand-planing. To joint the edge of a board, you can clamp it in the front vise. A board jack built into the bench will support the other end of long boards.

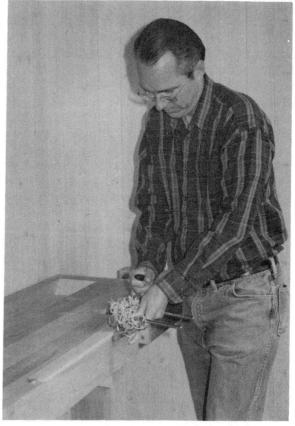

Illus. 1-1. A well designed and built workbench provides a stable working surface that makes it easier to perform many woodworking operations. Here I'm beading a board with a plane. The tail vise and dogs of the classic cabinetmaker's bench (described in Chapter 4) make it easy to hold even a thin, narrow board like this one.

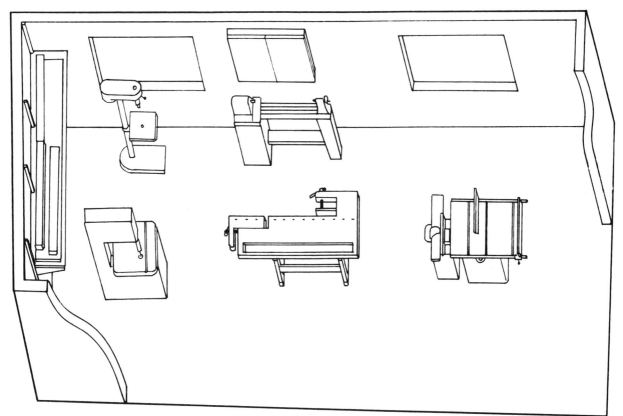

Illus. 1-2. Placing the workbench in the middle of the shop with other tools clustered around it makes it more versatile and improves the work flow.

To plane the face of a board, use bench stops or clamp the work between the tail vise and a bench dog. The tail vise is particularly useful when you are using a moulding plane or cutting rabbets, grooves, and dadoes with planes.

Mortising

A sturdy workbench makes mortising with a chisel easier. If you try to cut mortises on a flimsy table, the table will spring back with each blow of the mallet. This absorbs much of the energy from the blow, so the chisel doesn't cut very deep, and you must work harder. A thick bench top will direct more of the force from each blow where you want it, so you cut deeper with each blow. Holdfasts or other hold-down accessories help to keep the work in position as you cut mortises. When cutting tenons, you can clamp the work end-up in the front vise.

Dovetailing

The front vise is also useful for cutting dovetails. Nar-row boards can be clamped successfully in any type of front vise, but if you will be cutting dovetails on wide boards, then the shoulder vise is the best type of vise to use.

To chop the tails of a dovetail joint, you can stack the boards on a bench and clamp them with a holdfast or other type of hold-down.

Routing

A good workbench can also take the frustration out of many power-tool operations. It is particularly useful for routing. Most clamps will get in the way of a router eventually, so you end up stopping the cut to move the clamp. The tail-vise-and-bench-dog system can hold the work stationary without any projections above the work.

Sanding

A bench stop makes it easy to secure a piece of wood when belt-sanding. The belt sander tends to push the board rearwards. A bench stop at the end of the bench will prevent the board from moving.

Biscuit Joining

A flat bench top is important for many biscuit-joining operations. A bench stop or a hold-down can help to hold the work in position.

WORKBENCH MATERIALS

A workbench must be made from sturdy materials. The traditional favorites are closed-grain hardwoods such as maple and beech. Other tough hardwoods can also be used; oak is often used to build workbenches.

Softwoods such as pine and fir can also be used in workbench construction. Old Shaker workbenches were made of hardwood in the areas that receive the most wear, and pine in other areas. If you choose to use softwood with a design that calls for hardwood, increase the size of the parts to compensate for the difference in strength.

TYPES OF BENCH TOP

The bench top is the single most important part of a workbench. It must be strong, durable, flat, and stable. The tops of early workbenches were made of single, thick slabs of hardwood. A single slab of wood is strong and durable, but it is not stable and it is difficult to keep flat. As the wood dries, the bench top will cup. Old-time woodworkers had to periodically plane the bench top flat.

A laminated bench top is more stable than a single, thick slab. Laminated bench tops have been popular for over a hundred years and are still the type preferred by many woodworkers. There are two types of laminated tops: The uniform-thickness laminated top and the thickened-front laminated top (Illus. 1-3).

The uniform-thickness laminated top is usually about 2 inches thick and made from narrow strips of maple. You can buy ready-made uniform-thickness laminated tops. The ready-made tops are available from woodworking supply companies and industrial tool suppliers. They are a quick way to make a high-quality bench, because all you need to build is a base. You can make this type of bench top yourself, but since there are many glue joints, this can be a tedious procedure.

Most traditional cabinetmaker bench designs have a thickened-front laminated top. Since most of the heavy pounding takes place on the front surface of the bench top, this area needs to be the strongest. The thickened front edge provides extra strength in the area around the front dog holes. This type of top is usually made from wider boards. Since there are fewer glue joints, a thickened-front laminated top is easier to make yourself than the uniform-thickness bench top.

Even a laminated top will change shape slightly as its moisture content changes. The most stable bench tops are made from man-made materials (Illus. 1-4). Bench tops made from materials such as plywood, particleboard, and hardboard will remain flat even in conditions that would make a laminated, solid-wood top warp.

You can buy ready-made bench tops that are about

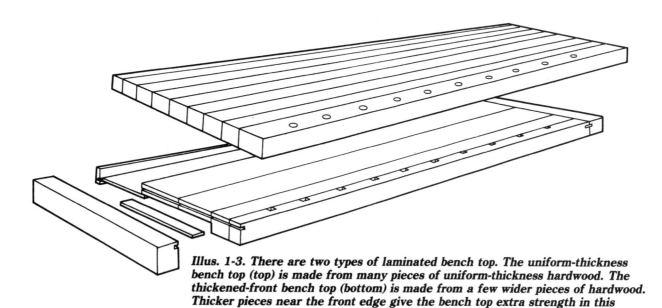

Illus. 1-3. There are two types of laminated bench top. The uniform-thickness bench top (top) is made from many pieces of uniform-thickness hardwood. The thickened-front bench top (bottom) is made from a few wider pieces of hardwood. Thicker pieces near the front edge give the bench top extra strength in this critical area.

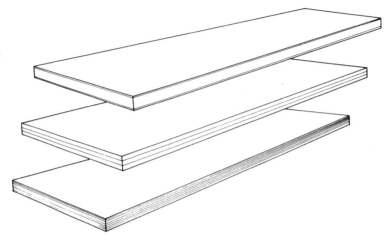

Illus. 1-4. Bench tops made from man-made materials are very stable. You can buy a bench top that has a particleboard core and a hardboard skin (top) or make a built-up top by gluing together several layers of particleboard (middle). A bench top made from special plywood with a thick face veneer (bottom) gives the bench a traditional look.

2 inches thick. They usually have a particleboard core and a hardboard skin. You can also make your own by building up the bench top from several layers of ¾-inch-thick particleboard. Tempered hardboard will endure much more abuse than particleboard, so for a really tough surface apply a skin of tempered hardboard to the particleboard core. If you would like a bench top as stable as one made with man-made materials, but one with a more traditional look, use hardwood plywood to make it. The one disadvantage of ordinary plywood is that the face veneer is fairly thin, so you can end up accidently cutting through it. Some commercial benches use special plywood that is designed for bench tops. It is 2-inch-thick plywood with ⅛-inch-thick face veneers.

Tool Trays

Many bench designs incorporate a tool tray in the bench top. The tool tray is a recessed area at the rear. Small tools can be placed in the tray; these tools will be below the top surface of the bench top, so that they will be accessible but out of the way.

It seems that woodworkers either love or hate tool trays. It all depends on how you work. The people who like tool trays enjoy having a handy place to keep a few tools while they work at the bench. Those who don't like them complain that the tool tray is just a dust catcher that uses up bench space. It is true that the tool tray is a dust catcher, but a dust ramp makes it easy to sweep out (Illus. 1-5).

BASE CONSTRUCTION

The main purpose of the base is to hold the bench top at a convenient height. It must be sturdy and stable. The best height for a workbench is a matter of debate

Illus. 1-5. Bench-top tool trays can provide a handy place to keep a few tools. A sloped board in the corner of the tool tray called a dust ramp allows you to sweep dust out of the tool tray.

among woodworkers. It depends on what kind of work you are doing and how tall you are. Generally, workbenches should be 30 to 36 inches high. Most commercial benches are about 35 inches high. I find it convenient to make my workbench the same height as my table saw. In a small shop, the workbench can double as an out-feed table for the table saw when you need to cut a large board.

There are three main types of base: the straight-leg, the sled-foot, and the post-and-panel base. Each of the three has advantages. The straight-leg base (Illus. 1-6) is easy to build. If you are making a narrow bench, the straight-leg design may not be the best, because the legs will be so close together that the bench could tip over. The sled-foot base (Illus. 1-7) makes a narrow bench more stable. The post-and-panel base (Illus. 1-8) is enclosed so that it can be used for storage.

Illus. 1-6. The straight-leg base is easy to build and is very strong.

Illus. 1-7. The sled-foot base makes a narrow bench more stable.

Illus. 1-8. The post-and-panel base is enclosed, so that it can be used for storage.

A workbench base must have sturdy joinery. The joints must be able to withstand the racking, pounding, and vibration associated with the woodworking operations that are performed on the bench. Another consideration when deciding on which joints to use on your workbench is that it may have to be moved to a different location after it is built. If you ever have to move your bench to another shop, it is more convenient if you can disassemble it. That is why bolts are often used to assemble the workbench base (Illus. 1-9). The bolt-and-cross-dowel fastener is a variation on the standard bolt. It allows you to install a bolt directly in the end

Illus. 1-9. Threaded fasteners strengthen the joints and make it possible to disassemble the bench for moving. The truss rod (left) provides the greatest strength. You can also use bolt-and-cross-dowel fasteners (middle) or standard bolts (right).

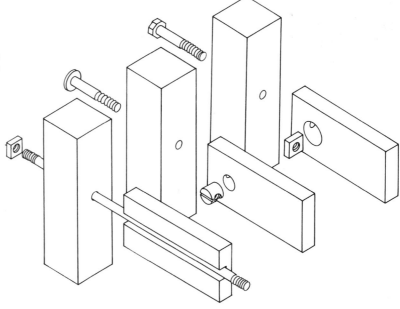

Illus. 1-10. If you prefer more traditional joinery, you can assemble the base using wedged mortise-and-tenon joints.

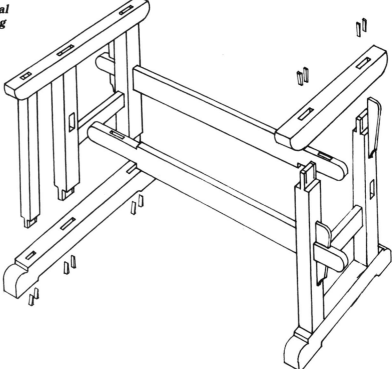

of a board. Truss rods are stronger than bolt- and cross-dowel fasteners, but you must cut a groove the full length of the stretchers to use them.

The mortise-and-tenon joint is the joint used most often in traditional designs. If you want to be able to disassemble the base easily, you can use wedged tusk tenons (Illus. 1-10) at key locations. A tusk tenon projects past the edge of the mating board.

TYPES OF WORKBENCH

Workbenches have been used since the dawn of woodworking. Early workbenches were simple, but by the 1600s they had developed into the form we recognize today. There have been gradual improvements in bench design over the years, but the basic design of a woodworking bench hasn't changed considerably for several hundred years. Although there are many variations, I will classify workbenches into two groups: joiner's benches and cabinetmaker's benches. Each is described below.

Joiner's Benches

In traditional woodworking terminology, a joiner is a woodworker who makes windows, doors, interior panelling, and other finish woodwork for buildings. A classic design for a joiner's bench (Illus. 1-11) was published in 1769 in a book called *The Art of the Woodworker* by Jacques-Andre Roubo. A joiner's bench is characterized by a large work surface and a simple, sturdy base. The joiner's bench design adapts well to modern materials, and it is especially useful for modern woodworkers who build large projects using power tools.

With the addition of vises and other bench accessories, a joiner's bench can also be used with hand tools. It is particularly useful when planing large boards and assembling large pieces of furniture.

A joiner's bench is a good choice if you want a versatile bench for many types of woodworking. It is also relatively simple and inexpensive to build, so it makes a good first bench.

Cabinetmaker's Benches

Cabinetmaker's benches are more specialized. They are designed to fit the needs of furniture makers who use mostly hand tools. Their designs are characterized by a smaller work surface and usually include a tool tray (Illus. 1-12). Vises are an important part of a cabinetmaker's bench. A front vise, a tail vise, and bench dog holes are essential.

Commercial Benches

There are several high-quality benches available in the United States. I will describe a few here to give you an idea of what is available.

Ulmia Hobbyist Workbench The Ulmia hobbyist workbench (Illus. 1-13) is a small joiner's bench. It has a $22\frac{5}{8} \times 59$-inch work surface with no tool tray. The top is laminated from solid beech. It is $3\frac{1}{2}$ inches thick

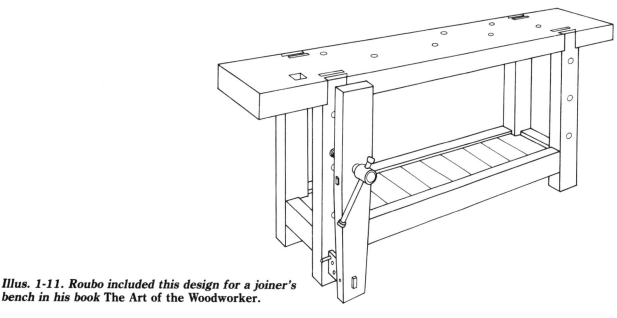

Illus. 1-11. Roubo included this design for a joiner's bench in his book The Art of the Woodworker.

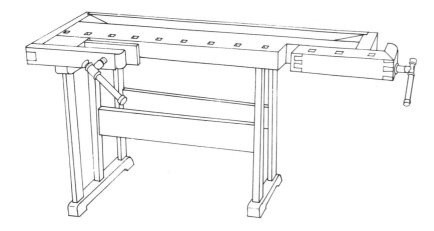

Illus. 1-12. This cabinetmaker's bench is based on a centuries-old Scandinavian design.

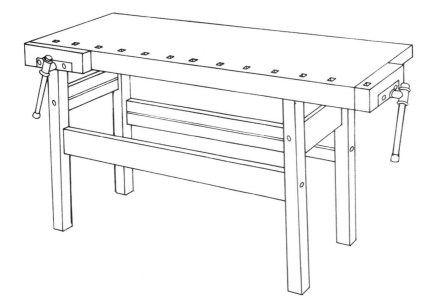

Illus. 1-13. Ulmia hobbyist work-bench.

at the edges and 1¾ inches thick in the middle. Both the front vise and the end vise have wooden faces. The top has a single row of square dog holes. The base is a simple straight-leg design assembled with truss rods.

Veritas Workbench This bench (Illus. 1-14) is a joiner's bench; it is 76 inches long and 26 inches wide. The top is made from 2-inch-thick beech plywood. There is a small tool tray in the middle. The base is a simple design assembled with truss rods.

This bench is designed to be used with the Veritas twin-screw vise, the Tucker vise, and the various other Veritas bench accessories. It has ¾-inch-diameter dog holes spaced every 7¾-inches. This bench is sold in kit form only. The kit includes all of the hardware and the 2-inch-thick plywood with dog holes predrilled. You

must supply the lumber to build the base. The vises are sold separately.

Ulmia Master Carver's Bench The Ulmia master carver's bench (Illus. 1-15) is a joiner's bench. It has a large 23½ × 72½-inch work surface made of laminated hardwood. The top is 3 inches thick at its edges and 1½ inches thick in the middle. There is a small removable tool tray. The top has two rows of square dog holes. The front vise is a wooden-face vise, and the end vise is a full-width wooden-face vise. The base is assembled with truss rods.

Sjoberg Swedish Workbench The Sjoberg Swedish workbench (Illus. 1-16) is a scaled-down version of the traditional Scandinavian cabinetmaker's bench. It is 26

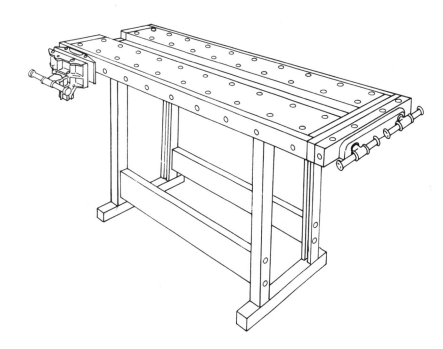

Illus. 1-15. Ulmia master carver's bench.

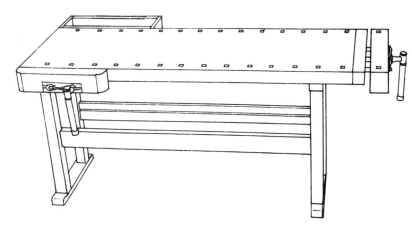

inches wide and 52 inches long. It has a laminated solid-lumber top with a rear tool tray. It includes a traditional end vise and a shoulder vise. It has one row of square dog holes. It is available with a simple sled-foot base or the cabinet base shown in Illus. 1-16.

Ulmia Cabinetmaker's Bench The Ulmia cabinet-maker's bench (Illus. 1-17) is a full-sized cabinetmak-er's bench with a wooden-face vise and a traditional end vise. The work surface is 16 inches wide and 63 inches long. A large tool tray at the rear adds 6 inches to the width. The top is laminated solid beech. It is 3 inches thick at the front edge and 1½ inches thick in the middle. There is a single row of square dog holes along the front edge. The base is assembled with truss rods.

Ulmia Ultimate Cabinetmaker's Bench This top-of-the-line cabinetmaker's bench (Illus. 1-18) has an 18¼ × 90⅛-inch work surface with a 6-inch-wide tool tray at the rear. The top is laminated from solid beech. It is 4 inches thick at the front edge and 2½ inches thick in the middle. There is a single row of square dog holes along the front edge. The post-and-panel base is assembled with truss rods. A large tilt-out drawer provides tool storage in the base.

BUILDING YOUR OWN BENCH

There are some advantages to building your own work-bench. First, commercial benches are fairly expensive, so you can save some money by building your own. Second, a workbench is a fun project that will also give

Illus. 1-16. Sjoberg Swedish workbench.

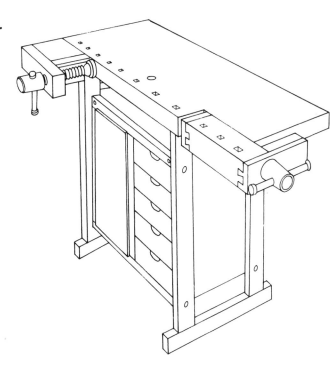

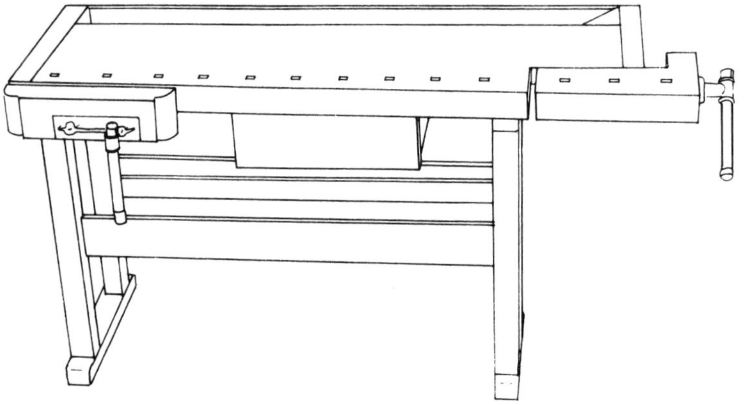

Illus. 1-17. Ulmia cabinetmaker's bench.

Illus. 1-18. Ulmia ultimate cabinetmaker's bench.

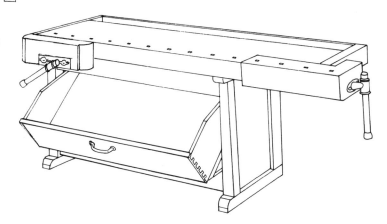

you valuable woodworking practice. Third, you can take pride in a bench that you built yourself.

In this book, I present plans and step-by-step in-structions for building two different workbenches with several variations. These benches are designed to use simple joinery, and are easy to build. They have all the features that can be found on workbenches, including the best features found on commercial benches.

In Chapter Three, I describe how to build joiner's benches. The joiner's bench (Illus. 1-19) is simple and inexpensive to build. A large work surface and a grid of round dog holes make it very versatile. It is a good bench for woodworkers who do a lot of power-tool work and who often build large projects. It is also very good for hand-planing large boards and all types of hand-tool operations. A laminated solid-wood top or a particleboard top can be used with a hardboard skin. Chapter Three also includes plans for a smaller version of the joiner's bench.

In Chapter Four, I describe how to build a classic cabinetmaker's bench (Illus. 1-20) based on a Scandinavian design. It includes a laminated solid-wood top, a tool tray, a shoulder vise, and a traditional tail vise. This design is best for those who want to specialize in fine furniture-making using traditional techniques. The shoulder vise is especially adapted to hold parts for hand-cutting dovetail joints.

SAFETY PROCEDURES

Take your time and have fun as you build your bench. Remember to work safely. To work safely, do the following:

1. Always follow the warnings and instructions that come with your tools.
2. Use the guards provided with your power tools.
3. Keep your hand tools sharp and in good condition.
4. Stay alert as you work.
5. Keep your hands away from the cutting edges of tools.
6. Use push sticks when necessary to keep your fingers away from the blade of the power tool.
7. Wear the proper safety equipment when woodworking. This includes safety glasses or goggles, hearing protectors, and a dust mask when you are sanding or doing other operations that produce a lot of dust.

Illus. 1-19. You will learn how to build this versatile joiner's bench in Chapter Three.

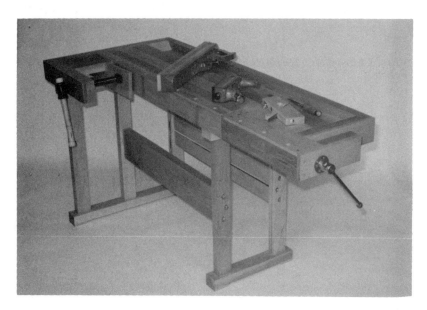

Illus. 1-20. You can build this classic cabinetmaker's bench by following the instructions given in Chapter Four.

CHAPTER TWO

Vises

ALTHOUGH MANY EARLY benches didn't have a vise, today most woodworkers consider a good vise an essential part of the workbench. Most workbenches are designed to accommodate vises, so the type of vise you choose often determines what the rest of the bench is like.

One of the best ways to upgrade an existing bench is to install a high-quality vise. There are several types to choose from. Some are made from cast iron, and others are mostly wood with some steel parts.

There are two main types of vises: front vises and tail vises. A front vise is used to hold boards on edge for planing or end up for operations such as cutting dovetails. The tail or end vise is used to hold a board flat on the bench top for planing the face. The tail vise is also useful for securing work when you are sanding or routing with power tools. Front and tail vises are described below.

FRONT VISES

There are two main types of front vise: iron vises and wooden vises. Below, I first discuss the different types of modern cast-iron vise, and then wooden front vises.

Iron Vises

Choose a high-quality cast-iron vise for the front vise. Record, Paramo, Jorgensen, Veritas, and other manufacturers make high-quality vises (Illus 2-1). Some vises are available with a quick-action screw release (Illus. 2-2 and 2-3). This feature allows you to slide the vise jaw to the approximate position quickly and then tighten it the rest of the way with the screw. This feature is handy, but not absolutely necessary.

There are three ways to mount a vise (Illus. 2-4). The simplest way is to mount it directly on the front edge of the bench. This is fine for many uses, but most

Illus. 2-1. Several companies make high-quality iron vises. The vise on the left is made by Yost. The vise in the middle is made by Record, and the vise on the right is made by Jorgensen.

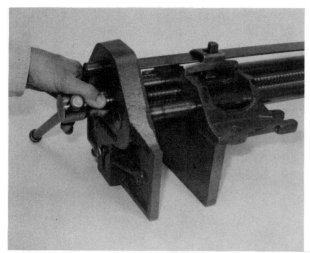

Illus. 2-2. *The quick-action screw release on this Record vise lets you move the vise jaw into position quickly. A lever releases the screw.*

Illus. 2-3. *This Jorgensen vise uses a gravity-operated roller nut. Turning the vise handle one-half turn counterclockwise disengages the screw, allowing you to slide the jaw to any position. Turning the vise handle clockwise reengages the screw.*

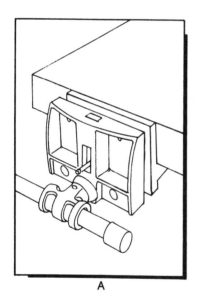

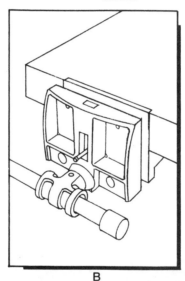

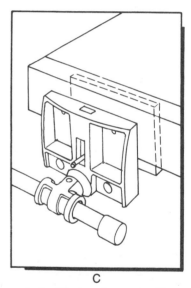

A   B   C

Illus. 2-4. *Three ways to mount a vise. A. Vise mounted with rear jaw extending past the front edge of the bench top. B. Vise mounted in notch so that rear jaw is flush with front edge of the bench top. C. Vise mounted with rear jaw behind front-edge strip of bench top.*

woodworkers find that it is better to have the inside or rear jaw flush with the front of the bench; this way, the front edge of the bench will help support the work and keep it steady. This is particularly important when the board is long.

When you are adding the vise after the bench is built, you can cut a notch in the front of the bench, as shown in B in Illus. 2-4. If you install the vise as you build the bench top, you can use the method shown in

C in Illus. 2-4. This is a very good way to install a vise. Use a sabre saw to cut the notch (Illus. 2-5).

The vise can be attached with lag screws, but bolts are better. A vise is subjected to a lot of pounding and twisting during its life. Lag screws will eventually pull out of the wood. Bolts will be more secure, and they can be retightened if they do loosen.

If the vise is designed for a bench top that is thicker than the one you are mounting it on, you will have to

add some spacers. Place the vise in position and mark the bolt-hole locations. Drill the bolt holes through the spacers and the bench top. Use a drill guide to keep the holes straight (Illus. 2-6). Countersink the top of the hole so that the bolt heads will be below the surface of the bench top. If you want to hide the bolt heads, countersink them deeper and plug the holes with dowels.

If you install the bolts before gluing on the top surface of the bench, you can hide the bolt heads as shown in Illus. 2-7. Use square-headed bolts. Drill the bolt holes, and then insert the bolts. Trace around the bolt heads, and then remove the bolts. Use a chisel to cut a square recess in the bench top. Insert the bolts and pound their heads into the recess until the top of the bolts are flush with the wood surface. Put the vise in place, put a lock washer over each bolt, and screw on the nuts. When you glue on the top surface, the bolt heads will be completely hidden.

Iron vise jaws can mar wood surfaces. The vise has screw holes in its jaw for attaching a wooden jaw pad. Cut the wooden jaw pad to fit your vise and attach it to the jaw with screws (Illus. 2-8). Jorgensen makes a magnetic jaw pad for its vises. It can be easily installed and removed (Illus. 2-9).

The Patternmaker's Vise The patternmaker's vise is unique because it allows you to tilt or rotate the work while it is clamped in the vise. This type of vise was originally made by the Emmert Company. A similar vise called the Tucker is currently being sold by the Veritas Tool Company (Illus. 2-10).

A patternmaker's vise is ideal for carvers. It is also useful for general woodworking. In its natural position, the vise operates like other quick-release vises, but with a few adjustments, you can pivot the front jaw to accommodate irregularly shaped parts or tilt and rotate the vise to place the part in its optimum working position (Illus. 2-11).

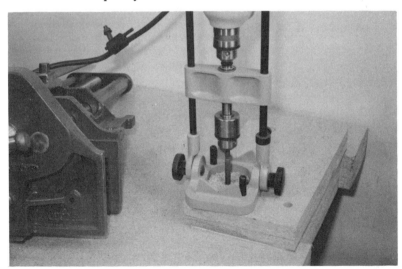

Illus. 2-6. You may need to add blocks to the underside of the bench top. Use a drill guide as you drill the bolt holes.

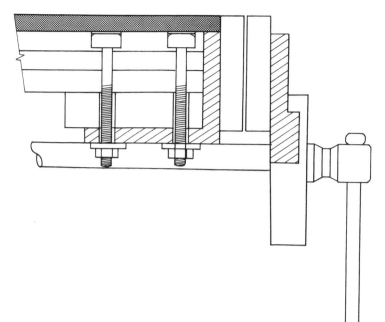

Illus. 2-7. If you countersink the bolt heads below the surface before applying the top skin of the bench top, the bolts will be completely hidden.

Illus. 2-8. Iron vise jaws can leave marks on the wood. To prevent damage to smooth wood surfaces, add a wooden jaw pad.

Illus. 2-9. The Jorgensen Company sells magnetic jaw pads for its vises. The pads are made of wood, and have a magnetic surface on their back and felt on their front. They prevent the iron jaws from damaging smooth wood surfaces. They can be installed and removed quickly, because they are held in place by magnets.

Illus. 2-10 (left). This patternmaker's vise is called a Tucker vise and is sold by Veritas Tool Company.

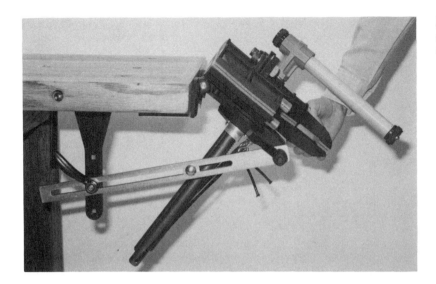

The patternmaker's vise has two sets of jaws. The large jaws are more useful. They can hold large parts securely. They also have dog holes that will accept ¾-inch-diameter round dogs, so you can clamp larger parts between dogs in the vise and dogs in the bench. You can rotate the vise to use the small jaws. They are useful for holding small parts (Illus. 2-12).

Illus. 2-12. The patternmaker's vise has two sets of jaws. The small jaws are useful for holding small parts.

Bench thickness is critical when you install a Tucker vise. If the bench top is too thin, add a spacer (Illus. 2-13). If the top is too thick, you must cut a recess in the bottom of the bench to accommodate the vise. You

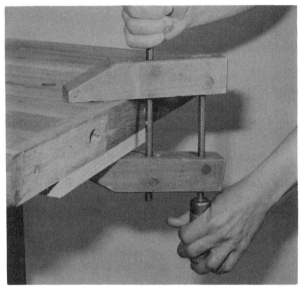

Illus. 2-13. If the bench top is too thin, you must add a spacer to install a Tucker vise.

must also cut a groove in the edge of the bench top. This groove helps keep the vise in alignment. Use a router to cut the groove (Illus. 2-14).

Wooden Vises

Originally, all vises were made of wood. The earliest ones even had wooden screws. There were several types commonly used on old benches. The *leg vise* (Illus. 2-15) was common because it was easy to build, but it had some limitations. Its jaw was not very wide, and a horizontal beam at the bottom had to be adjusted for different thicknesses. The *twin-screw face vise* (Illus. 2-16) was an improvement. It was capable of clamping wide boards end-up for dovetailing and similar operations. In fact, the design still survives today in the Veritas twin-screw vise. This vise is described more thoroughly in the tail vise section. After many years of evolution, the *single-screw face vise* and the *shoulder vise* became the most widely used types of wooden vise. These vises are described below.

Illus. 2-14. Use a router to cut the groove that aligns the Tucker vise-mounting plate.

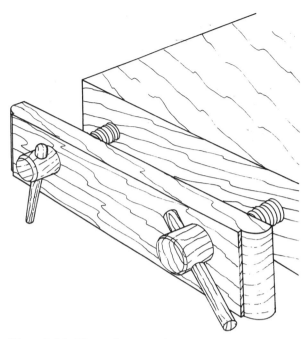

Illus. 2-16. The twin-screw face vise clamps wider boards than the leg vise.

Single-Screw Face Vise The single-screw face vise (Illus. 2-17) originally had a wooden screw and guide. Today, most have metal screws and guides, but they retain the wooden jaws. This vise is one of the best types of wooden vise for clamping a board edge-up for planing (Illus. 2-18). It is also useful for most general work. It can be used to clamp narrow boards end-up for dovetailing, but it won't hold wide boards securely.

Illus. 2-15. The leg vise was an early form of wooden vise.

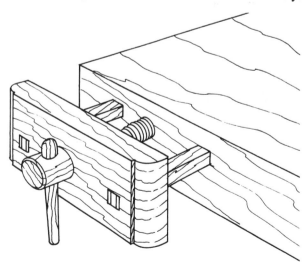

Illus. 2-17. The single-screw face vise is the most popular wooden vise today. Originally, the screw and guides were made of wood, but today most of these vises have metal guides and screws.

Illus. 2-18. This modern single-screw face vise works well for clamping boards edge-up for planing.

You can overcome this shortcoming by clamping one edge of a wide board in the vise and using a bar clamp to secure the other edge to the bench (Illus. 2-19). If you don't have to clamp wide boards very often, this is probably the best all-around vise.

You can buy the steel hardware to make a single-screw face vise from most woodworking supply companies. The simplest way to make this type of vise is to buy hardware that includes guide rods (Illus. 2-20). Then all you have to do is make a jaw out of thick hardwood and mount the hardware on the bench. You can also buy the vise screw separately and make your own guide rods from pipe or steel rod.

Some vise hardware comes with an iron guide plate that bolts to the underside of the bench top. This makes it easy to make and install the vise. All you have to do is make the wooden jaw and then bolt the hardware to the bench. If you use this type of hardware, the size

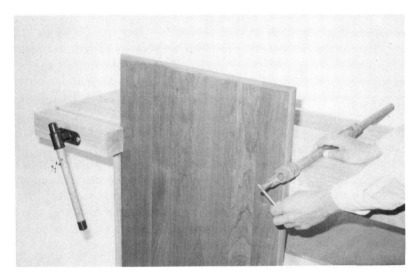

Illus. 2-19. The guides on the single-screw face vise get in the way when you clamp a wide board. You can overcome this problem by using a bar clamp to secure the free end of the work.

Illus. 2-20. This vise uses commercial steel hardware.

of the vise is limited by the guide rod spacing. If you make your own guides, you can space the guide rods as far apart as you like to make an extra-long vise. Make the guides from thick hardwood and drill holes for the guide rods and screw. Mount the guides to the underside of the bench top (Illus. 2-21).

Illus. 2-21. The steel rods fit into hardwood guides mounted on the underside of the bench.

Shoulder Vise The shoulder vise (Illus. 2-22) is a better choice than a front vise if you will be doing a lot of hand dovetailing. It can hold a wide board end up. It is also good for clamping irregularly shaped parts, because the front jaw pivots on the vise screw. It is more complicated to build because it is an integral part of the bench. See Chapter Four for complete plans and instructions for building a shoulder vise. You can buy the vise screw (Illus. 2-23) from most woodworking supply companies.

TAIL VISE

The tail vise is used to clamp a board between a dog on the vise and a dog in the bench top. To use an end or tail vise, you will need bench dogs. These are discussed in Chapter Three.

The simplest way to add a tail vise is to attach a second iron woodworking vise to the end of the bench. Most woodworking vises have a retractable dog for this purpose (Illus. 2-24). This vise doesn't need to be recessed into the edge.

Many times it is useful to be able to clamp a narrow board with the end vise and have the edge of the board slightly overhang the edge of the bench. Using a router to cut a decorative edge on a board is one example where this setup works well. For this reason, it is good to have the dog holes close to the front edge of the bench. The dog holes must line up with the dog in the middle of the vise, so, in this case, with a smaller vise you can place the dog holes closer to the edge. A small vise isn't a disadvantage in the tail vise position. Because it only needs to extend as far as the spacing between the dog holes, you can simply drill the dog holes closer together when you use a small end vise.

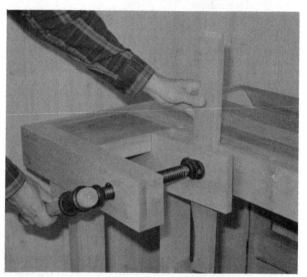

Illus. 2-22. The shoulder vise is a traditional wood vise that is still very popular today. It is useful for clamping wide boards and irregularly shaped boards.

Illus. 2-23. You can buy a vise screw like this one to use when building a shoulder vise.

Illus. 2-24. Many iron vises have a retractable dog. This makes it possible to use them as tail vises.

Two-Dog Tail Vise

A tail vise is very good for securing work for routing operations, because the dogs won't interfere with the router like clamps can. For holding large parts, it is better to have two dogs in the vise jaw.

The Tucker vise works well in the tail vise position. Because it has two dog holes in the jaw, you can drill two rows of dog holes in the bench. This makes it possible to hold large parts securely with four dogs (Illus. 2-25). Since you can adjust the angle of the vise jaw, you can even clamp irregularly shaped work between dogs.

You can make a wooden tail vise with two dog holes. It is basically the same as the wooden face vise described earlier. Make an extra-long vise jaw and drill a dog hole at each end (Illus. 2-26). This system works fine as long as you always clamp a board that is wide enough to reach both dogs. However, since the dog holes are not aligned with the vise screw, the vise will tend to twist when you only use one of the dogs. This means that you can't use as much force, and it can eventually damage the vise. You can partially solve this problem by spacing the guide rods as far apart as possible.

Twin-Screw Tail Vise

The twin-screw vise (Illus. 2-27) allows you to clamp the work with more than one dog or exert full pressure with only one dog in the jaw. It has two screws that are connected with a chain-and-sprocket drive, so that turning one handle rotates both screws (Illus. 2-28). This is a very good vise if you frequently need to clamp large parts to the workbench. You can buy the vise hardware from Veritas Tool Company. You can adjust the vise to fit any size bench. To install the vise, you have to make an end cap for the bench and a vise jaw from hardwood.

This vise is more versatile than most end vises; a release pin allows you to disengage the chain drive so that you can angle the jaw slightly to compensate for irregularly shaped parts. It can also be used like a shoulder vise to clamp boards end up. The width of the board is limited by the distance between the screws, usually around 16 inches.

Illus. 2-25. The Tucker vise works well as a tail vise, because its two dog holes make it possible to hold wide boards securely.

Illus. 2-26. A wooden single-screw face vise can be mounted on the end of the bench to serve as a tail vise. If you drill two dog holes in the vise jaw, you can use it to clamp wide boards.

Illus. 2-27. This modern version of the twin-screw vise uses hardware made by Veritas Tool Company. It is very useful as a tail vise.

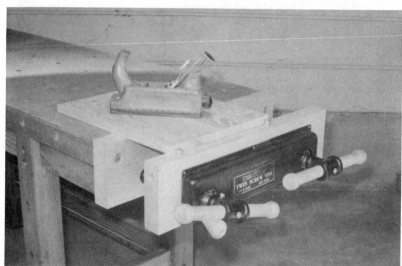

Illus. 2-28. A chain-and-sprocket drive connects the twin screws on this Veritas twin-screw vise.

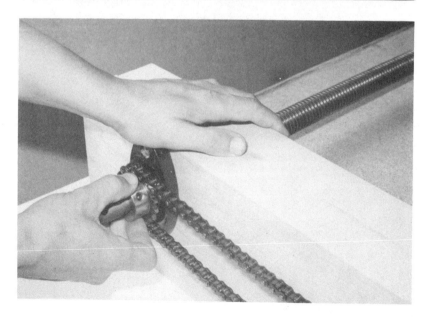

Traditional Tail Vise

The traditional tail vise was developed especially for holding boards for hand planing (Illus. 2-29). It is an integral part of the bench. This type of vise is more complicated to install than the other tail vises. Traditionally, it has had a wooden guide system. However, you can also buy steel guides. See Chapter Four for complete plans and step-by-step instructions for building a traditional tail vise.

Illus. 2-29. This traditional tail vise is an integral part of the bench top. Chapter Four includes step-by-step instructions for building this traditional tail vise.

CHAPTER THREE
The Joiner's Bench

I N THIS CHAPTER I describe how to build a sturdy and versatile joiner's bench (Illus. 3-1). Anyone with basic woodworking skills should be able to build this bench. This bench is based on a design originally published in 1769 in the book *The Art of the Woodworker*, by André Roubo, but I have adapted the design to use modern materials. The original had a massive solid-wood top. Solid wood has been the traditional choice for workbench tops, but modern materials such as laminated hardwood, plywood, particleboard, and tempered hardboard are much more stable.

The base of this joiner's bench is made of standard dimensional lumber (Illus. 3-2). For the bench top, use either laminated solid wood or particleboard and tempered hardboard. You should be able to buy most of the materials at a local lumberyard. The laminated solid-wood bench top will probably have to be bought from a mail-order woodworking supply company or a local industrial supply dealer.

A wide range of accessories can be used with this bench. It is designed to accept any type of vise, including a twin-screw end vise. The vise hardware can be bought by mail from several different woodworking supply companies. If you drill a grid of ¾-inch-diameter dog holes, you will be able to use a variety of clamps with this bench (Illus. 3-3). Chapter Five describes many hold-downs, stops, and other bench accessories that will fit in the ¾-inch dog holes.

The base is assembled with truss rods. Truss-rod joints are easy to make and very strong. The bench

Illus. 3-1. You can build this sturdy and versatile joiner's bench by following the directions in this chapter.

Illus. 3-2. The base of this joiner's bench is made of standard dimensional lumber. You can use either a factory-made laminated solid-wood bench top or one that you make yourself out of particleboard and tempered hardboard.

Illus. 3-3. If you drill a grid of ¾-inch dog holes in the bench top, you will be able to use many different dogs, hold-downs, stops, and other bench accessories.

Illus. 3-4. You can use the same base for any bench top that ranges in size from 24 × 48 inches to 24 × 72 inches. This bench has a 24 × 58-inch factory-made laminated solid-wood bench top.

legs are flush with the front edge of the bench top. This gives the bench a wide footprint for stability, and it makes it possible to use the legs as board jacks to support long boards in the front vise.

You can adapt the bench for specific needs by vary-ing the size of the bench top. Any size top from 24 × 48 inches to 24 × 72 inches can be used. The same base dimensions can be used for any top that falls within this size range (Illus. 3-4). Illus. 3-5—3-10 show the plans for the joiner's bench.

Joiner's Bench Bill of Materials

	Description	Size (Inches)	Material	Number Required
A	Top	Variable (See Text)	Variable (See Text)	1
B	Legs	$3\frac{1}{2} \times 3\frac{1}{2} \times 32$	4×4 Fir	4
C	End Stretchers	$1\frac{1}{2} \times 3\frac{1}{2} \times 17$	2×4 Fir	4
D	Side Stretchers	$1\frac{1}{2} \times 3\frac{1}{2} \times 38$	2×4 Fir	4

Joiner's Bench Hardware List

Number Required	Description
4	$\frac{3}{8}''$ Threaded Rod (6-Foot Lengths)
16	$\frac{3}{8}''$ Nuts
16	$\frac{3}{8}''$ Washers
10	$\frac{1}{4}'' \times 2\frac{1}{2}''$ Lag Bolts
10	$\frac{1}{4}''$ Washers
32	$\frac{3}{8}'' \times 1\frac{1}{2}''$ Dowels

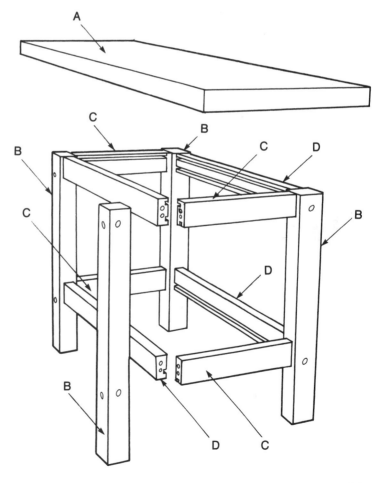

Illus. 3-5. Exploded view of joiner's bench.

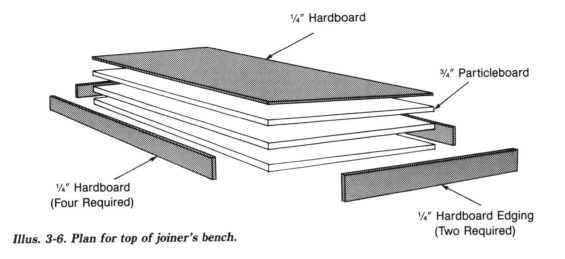

¼″ Hardboard

¾″ Particleboard

¼″ Hardboard
(Four Required)

¼″ Hardboard Edging
(Two Required)

Illus. 3-6. Plan for top of joiner's bench.

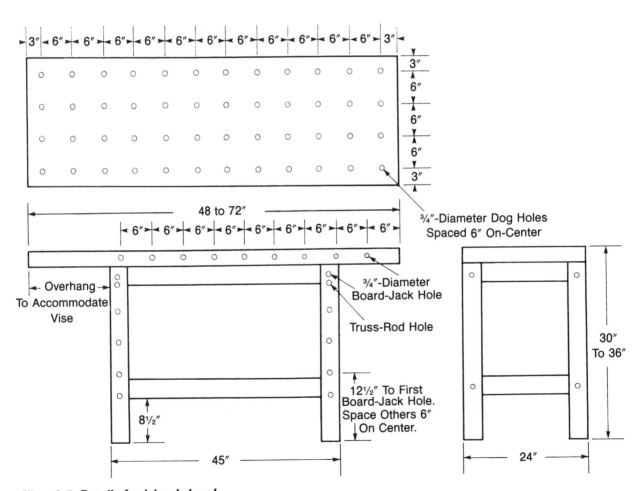

3″ 6″ 6″ 6″ 6″ 6″ 6″ 6″ 6″ 6″ 6″ 3″

3″
6″
6″
6″
3″

48 to 72″

6″ 6″ 6″ 6″ 6″ 6″ 6″ 6″

¾″-Diameter Dog Holes
Spaced 6″ On-Center

←Overhang→
To Accommodate
Vise

¾″-Diameter
Board-Jack Hole

Truss-Rod Hole

12½″ To First
Board-Jack Hole.
Space Others 6″
On Center.

8½″

45″

30″
To 36″

24″

Illus. 3-7. Details for joiner's bench.

Illus. 3-8. Details for joiner's bench.

Part C · · · Part D

Upper Stretchers

3/8"
3/8"

Groove For Top Hold-Down Clip

3/8"
1"
3/8"-Diameter Dowel Holes
1

3/8"-Diameter Dowel Holes

1 1/4" · 3/8"
3/8"

1 1/4" · 3/8"

3/8"-Diameter Steel Truss Rod

3/8"

3/8"-Diameter Steel Truss Rod

Lower Stretchers

3/8"-Diameter Dowel Holes

3/8"-Diameter Dowel Holes

1 1/4" · 3/8"
3/8"

1 1/4"
3/8"

3/8"
3/8"

3/8"-Diameter Steel Truss Rod

3/8"-Diameter Steel Truss Rod

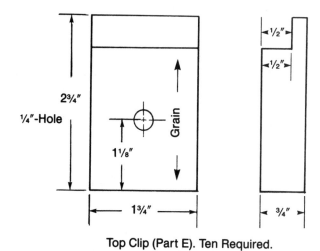

2 3/4"

1/4"-Hole

1 1/8"

Grain

1/2"

1/2"

1 3/4"

3/4"

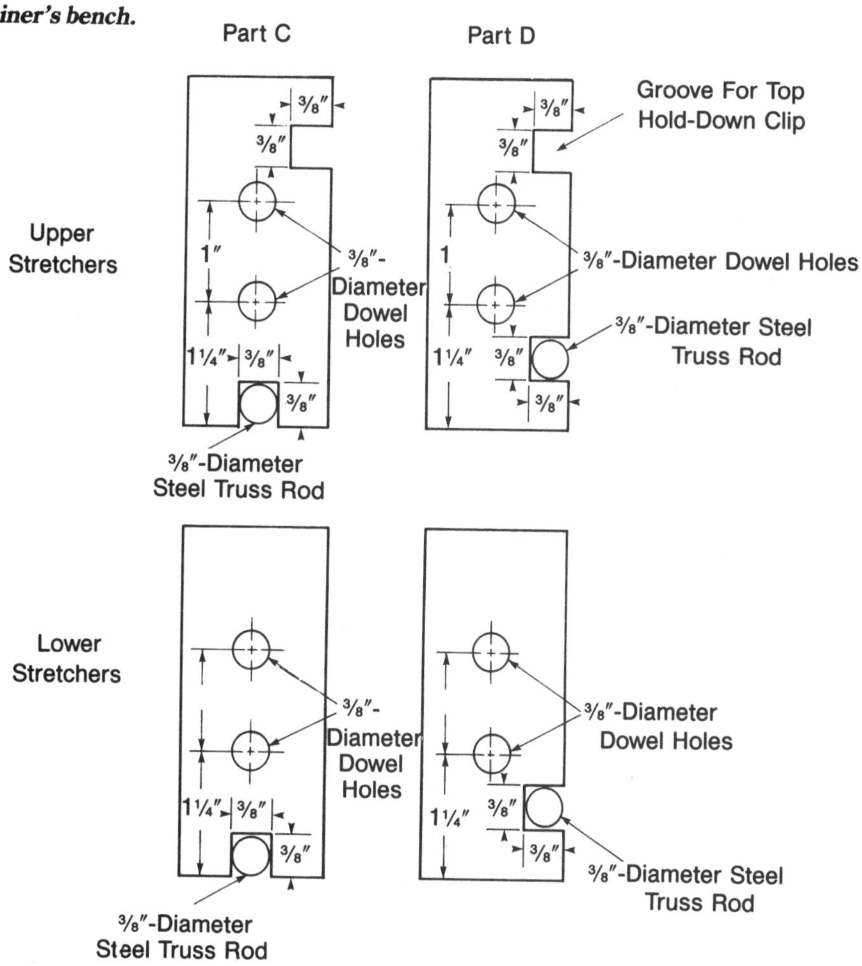

Top Clip (Part E). Ten Required.

Illus. 3-9. Details for joiner's bench.

31

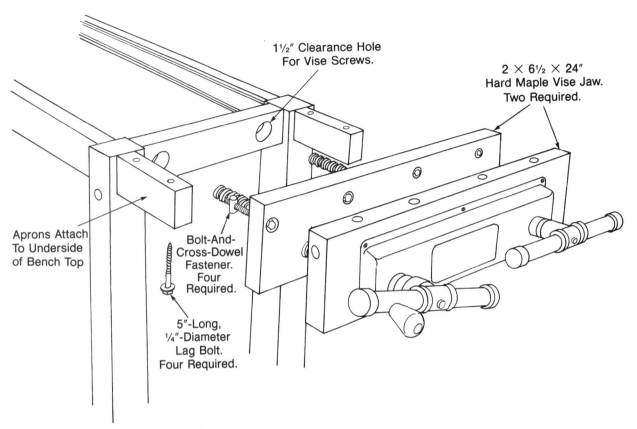

Illus. 3-10. Details for joiner's bench.

Labels in illustration:
- 1½" Clearance Hole For Vise Screws.
- 2 × 6½ × 24" Hard Maple Vise Jaw. Two Required.
- Aprons Attach To Underside of Bench Top
- Bolt-And-Cross-Dowel Fastener. Four Required.
- 5"-Long, ¼"-Diameter Lag Bolt. Four Required.

The maple bench top comes prefinished. You can leave the hardboard bench top unfinished or apply a coat of penetrating oil. Finish the base as you like. The simplest finish is a penetrating oil finish.

Following are instructions on how to build a joiner's bench. In Chapter Six, I describe how to use it.

BUILDING THE BENCH TOP

If you choose to use a laminated solid-wood bench top, I recommend that you buy a ready-made laminated hard-maple bench top from a local industrial supply company or a mail-order woodworking supply company. If using a ready-made bench top, skip the rest of this section and start by drilling the dog holes.

If you choose to build the top from particleboard and hardboard, you will need two full sheets of ¾-inch-thick particleboard and one-half sheet of ¼-inch-thick tempered hardboard. If you don't have a way to transport and cut full 4 × 8-foot sheets, have the lumberyard cut the parts for the top to size.

You can cut the parts for the bench top yourself with a portable circular saw. Use one edge of the sheet of tempered hardboard as a guide for the saw. Measure the distance from the edge of the blade to the edge of the saw base; this is the guide offset measurement.

After you have marked the location of the cut on the first sheet of particleboard, measure from that mark and make another mark at the guide offset measurement. Make a mark at both ends of the cut. Clamp the hardboard to the particleboard so that the edge lines up with the guide marks. Hold the edge of the saw base against the edge of the hardboard as you make the cut.

Repeat this procedure for all of the cuts. Finally, cut the hardboard using one of the factory edges of a piece of particleboard as a guide (Illus. 3-11).

Now, you are ready to laminate the top. Nails are used to assemble the top. If you will be drilling dog holes in the top, lay out the positions of these holes now so that you won't drive a nail where you will need to drill a hole later.

Place the first piece of particleboard on a flat surface and spread an even coat of aliphatic resin glue over its entire surface. You can use a small paint roller to produce an even coat (Illus. 3-12). Apply a coat of glue to the mating surface of the second piece of particleboard,

Illus. 3-11. Use a portable circular saw to cut the parts for the bench top.

Illus. 3-12. Apply an even coat of glue to the particleboard. You can use a small paint roller to spread the glue.

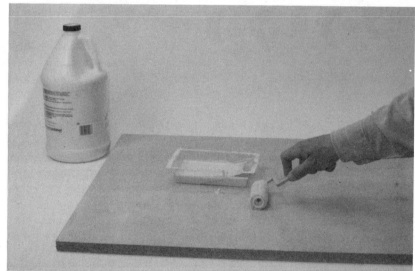

Illus. 3-13. Drive nails about 6 inches apart to hold the boards together as the glue dries. If you will be drilling dog holes in the bench top, be sure to mark the location of the dog holes so that you don't drive a nail where the drill will hit it.

and then place it on top of the first piece. Align the edges of the two pieces, and then drive some finishing nails to hold them together. Drive more nails about six inches apart to apply even pressure to the glue joint (Illus. 3-13). Repeat the gluing and nailing procedure for the third piece of particleboard.

If you want to hide the vise mounting bolts, mount the vises before you apply the hardboard skin. See Chapter Two for details.

Next, glue on the hardboard edge strips. If you don't mind some visible nail holes, you can use aliphatic glue to attach the hardboard. Apply a heavy coat to both mating surfaces and put the hardboard in position. Use a few finishing nails to hold it in place while the glue dries. You can also use bar clamps to apply even pressure to the edge strips as the glue dries.

To avoid any visible nail holes, apply the hardboard with contact cement. Apply two coats of contact cement to both mating surfaces. When the contact cement is dry, align the parts and press the hardboard in place (Illus. 3-14). Trim the edge strips so that they are flush with the top surface of the particleboard.

Now, apply the hardboard skin to the top face. Cut the skin about ¼ inch larger than the finished size. This allows you to trim it to fit after it is in place.

Apply two coats of contact cement to both mating surfaces. After the contact cement is dry, have someone help you put the skin in place. The contact cement will stick as soon as the two surfaces touch, so be sure to align the edges carefully.

To ensure a good bond, apply pressure to the surface with a hammer or laminate roller. To use a hammer, place a piece of scrap wood on the surface of the skin and hit the scrap with the hammer. Move the scrap slightly and hit it again. Continue this way until you have covered the entire surface. A laminate roller is designed to apply uniform pressure to contact cement. Press down hard on the roller and roll it over the entire surface several times.

Next, use a router to trim the edges flush. Install a laminate-trimming bit in the router. This bit has a pilot that rubs against the edge to guide the cut (Illus. 3-15).

DRILLING THE DOG HOLES

If you want to add dog holes to your bench, now is the best time to drill them. If you will be using a Tucker vise, space the holes 7¾ inches apart. Otherwise, use the 6-inch spacing shown in Illus. 3-7. You can choose how many holes you drill. One row across the front is traditional, but I prefer the full grid shown in Illus. 3-7. The grid of dog holes lets you take full advantage of a wide variety of hold-downs and other bench accessories.

Lay out the hole positions accurately. In Chapter Five, I show you how to make some bench accessories that must fit in two dog holes at once. If the holes aren't spaced accurately, you won't be able to use these accessories.

You will need a ¾-inch drill bit and a drill guide (Illus. 3-16) to space the holes accurately. The drill guide has a base that rests on the surface and guides that keep the drill square with the work. It is important to keep all of the dog holes straight, so don't try to drill them freehand.

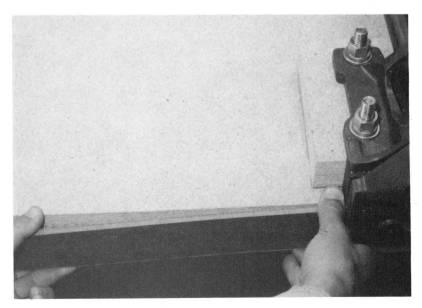

Illus. 3-14. You can apply the edging strip with contact cement. Align the parts carefully, because once they touch, they cannot be moved.

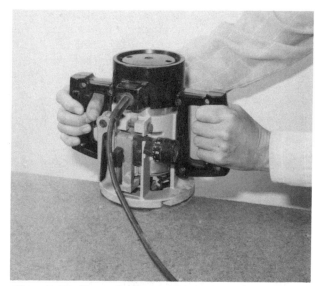

Illus. 3-15. Use a laminate-trimming bit in a router to trim the edges of the top.

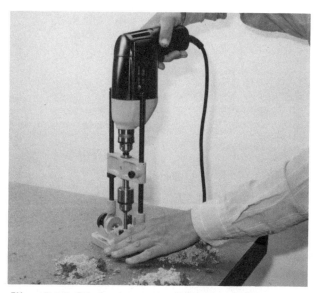

Illus. 3-16. It's important to drill the dog holes straight, so don't try to drill them freehand. Use a drill guide that keeps the drill square with the work.

Any type of ¾-inch wood bit can be used to drill the dog holes. Even an inexpensive spade bit will work. If you are drilling in particleboard, be prepared to sharpen the bit several times. Particleboard will dull the bit rapidly. Drill through from the top until the center spur of the bit breaks through the bottom face. After you have drilled all of the holes this far, turn the bench top over and finish the holes.

BUILDING THE BASE

The base is made from dimensional lumber. Use 4 × 4-inch fir for the legs and 2 × 4-inch fir for the stretchers. Cut the parts to length, and then cut the joints in the legs.

Next, cut the grooves in the stretchers for the truss rods. Use a router (Illus. 3-17) or a table saw with a dado blade to make this groove. The groove can also be made with a plow plane. Sand the legs and stretchers and round the edges slightly.

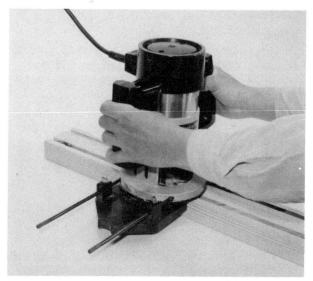

Illus. 3-17. Truss rods fit in grooves cut in the stretchers. You can cut the grooves with a router.

Carefully lay out the locations of the truss-rod holes in the legs (Illus. 3-18), and then drill the holes. Use a drill guide to keep the holes straight. Drill the 1-inch-diameter × ½-inch-deep countersinking holes

Illus. 3-18. After you have cut the grooves in the stretchers, lay out the locations of the truss-rod holes in the legs.

first (Illus. 3-19). After you have drilled all the countersinking holes, drill the $^{13}/_{32}$-inch holes completely through the legs (Illus. 3-20).

Next, drill the dowel holes. The dowels keep the parts in alignment as you assemble the base and keep the stretchers from twisting later. The truss rods hold the joints together and provide most of the strength.

Use a dowelling jig to drill two $^{3}/_{8}$-inch-diameter dowel holes in the ends of all the stretchers (Illus. 3-21). Next, lay out the position of the dowel holes in the legs. You can measure their locations or use dowel centers to locate them. Dowel centers are metal plugs that fit in the dowel holes in the stretchers. They have a sharp spur in the center that will make the center of the hole on the leg. Temporarily assemble each joint with the dowel centers in place. Then disassemble the joint and drill the hole using the mark from the dowel center as the center mark of the hole. The legs are too large to fit a dowelling jig, so use a drill guide to keep the holes straight (Illus. 3-22).

If you want board-jack holes in the front legs, drill them now. Use a $^{3}/_{4}$-inch drill bit and guide the drill with a drill guide (Illus. 3-23).

You can round over the corners of the legs, if you like. Use an edge-rounding bit in the router (Illus.

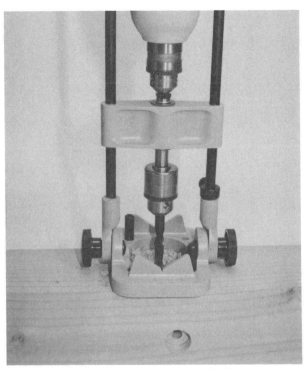

Illus. 3-20. After you have drilled all the countersinking holes, drill the $^{13}/_{32}$-inch truss-rod holes.

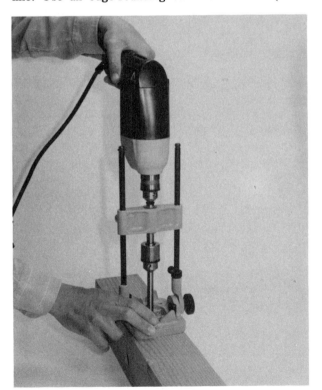

Illus. 3-19. The truss-rod nuts fit into one-inch-diameter countersinking holes in the legs. Drill these $^{1}/_{2}$-inch-deep countersinking holes first.

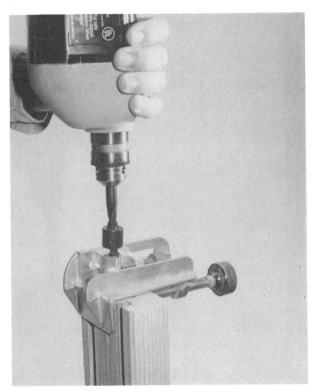

Illus. 3-21. Dowels keep the parts in alignment. Use a dowelling jig to drill the dowel holes in the ends of the stretchers.

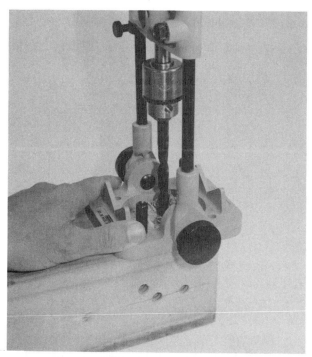

Illus. 3-22. *The legs are too large for the dowelling jig, so use a drill guide to drill the dowel holes in them.*

Illus. 3-24. *You can round over the corners of the legs with a router and an edge-rounding bit.*

Illus. 3-23. *If you want to include board-jack holes in the front legs, drill the ¾-inch-diameter holes before assembling the base.*

3-24). The bit I used has a pilot that guides the cut, so there is no need to use a router fence.

Now, assemble the base. Don't use any glue on the joints. That way, you can disassemble the base later if you need to move it. The truss rods will provide plenty of strength without glue. Insert the ⅜-inch dowels into the dowel holes in the stretchers. I used ready-made

dowels that are precut to 1½-inch lengths, but you can cut your own from ⅜-inch dowel if you like.

Next, cut the truss rods to their approximate lengths. The truss rods are made from ⅜-inch threaded rod. You will have to buy four 6-foot-long pieces. You will be able to get one long truss rod and one short truss rod from each 6-foot length. After cutting the rod, you will have to file the threads at the cut end before you can attach a nut.

Attach the end stretchers first. Place the truss rods in their grooves and assemble the joints (Illus. 3-25).

Illus. 3-25. *Place the truss rods in their grooves to begin assembling the base.*

Install washers and thread on the nuts (Illus. 3-26). Tighten the nuts with a socket wrench (Illus. 3-27).

After all the end stretchers are attached to the legs, install the front and rear stretchers (Illus. 3-28). When all of the nuts are tight, use a hacksaw to trim off the excess threaded rod.

Illus. 3-26. Push the truss rod through the hole in the leg and secure it with a washer and nut.

Illus. 3-27. Use a socket wrench to tighten the nuts.

Illus. 3-28. After all the end stretchers are attached to the legs, install the front and rear stretchers using the same procedure.

ATTACHING THE BENCH TOP TO THE BASE

The bench top is attached to the base with wooden clips held in place with ¼-inch lag screws. The clips fit into the top groove in the upper stretchers. Attaching the top this way allows it to move slightly as it shrinks and swells with changes in humidity. This is especially important if you are using a laminated maple top. If you were to bolt the bench top directly to the base, it would split when it shrinks as the wood dries out.

To make the clips, cut a ¾-inch-thick board 2¾ inches wide and about 16 inches long. Next, cut a ½ × ½-inch rabbet along one edge. Cut the rabbet on a table saw or with a router or rabbet plane. After the rabbet has been cut, cut the clips to size. Drill a ¼-inch hole in the clips. They are now ready to be used.

Turn the bench top upside down and put the base in place. Put the clips in place. Use three clips across the front and back and two on the ends. Space them equally and position them between the dog holes. Use a 7/32-inch drill bit to drill pilot holes for the ¼-inch lag screws. Hold the clip in place and place the drill bit in the hole you have already drilled in the clip; then drill the pilot hole into the bench top. To ensure that you won't drill completely through the bench top, use a drill stop or wrap some masking tape around the drill bit to indicate when to stop drilling.

The lag screws will be easier to install if you rub beeswax on the threads before you drive them. Now, put a washer on the lag screws and drive them into the hole. Use a socket wrench to tighten the lag screws (Illus 3-29).

INSTALLING A VERITAS TWIN-SCREW VISE

The Veritas twin-screw vise works well as a tail vise on this bench. Consult the directions that come with the vise hardware for complete instructions. Here I give an overview of how to install this vise.

You will need two maple vise jaws in addition to the vise hardware. The jaws are 2 inches thick × 6½ inches wide × 24 inches long. The inside jaw attaches to the end of the bench with bolts and cross dowels that are supplied with the vise hardware (Illus. 3-10). You will have to drill bolt holes in the end of the bench top and intersecting cross-dowel holes in the underside of the bench top. Insert the cross dowels and bolt the inside jaw to the bench top (Illus. 3-30).

Illus. 3-30. The inside jaw of the twin-screw vise attaches to the end of the bench top with bolt-and-cross-dowel fasteners.

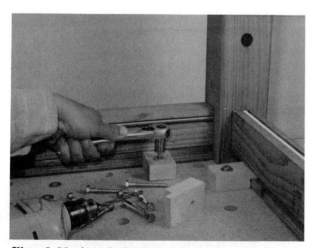

Illus. 3-29. Attach the top with wooden clips that fit into the grooves in the top stretchers. Attaching the top this way allows for dimensional change due to changes in the wood's moisture content.

Illus. 3-31. If you use a twin-screw vise, you must drill two holes in the stretcher that align with the vise screws.

The base has to be modified slightly to accept the vise hardware. You won't be able to close the vise fully unless you drill two holes in the stretcher that align with the vise screws (Illus. 3-31). You can also add the optional aprons shown in Illus. 3-10 to hide the vise screws. These aprons are strictly decorative. They are not necessary for strength, so you can omit them if you like.

Now, install the front jaw and the rest of the vise hardware (Illus. 3-32). The vise screws are connected with a chain. You will have to adjust the length of the

Illus. 3-32. *After the inside jaw has been attached, you can install the front jaw and the rest of the vise hardware.*

chain to fit the 14½-inch screw spacing needed for this bench. The instructions that come with the vise hardware will describe how to shorten the chain. Place the chain on the sprockets and install the sprockets on the

Illus. 3-33. *Next, install the chain sprockets and tighten the setscrews.*

shafts (Illus. 3-33). Next, attach the handle tees by driving in the spring pins (Illus. 3-34).

When you have completed the vise assembly, close the jaws and lay out the dog-hole locations. Align the dog holes in the vise jaw with the ones you have already drilled in the bench top. Now, drill the dog holes with a ¾-inch drill bit. Use a drill guide to keep the drill straight (Illus. 3-35).

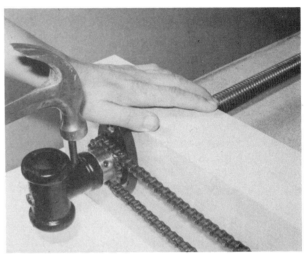

Illus. 3-34. *Attach the handle tees by driving in the spring pins.*

Illus. 3-35. *Drill dog holes in the vise jaw that align with the dog holes in the bench top.*

CHAPTER FOUR

The Classic Cabinetmaker's Workbench

THIS WORKBENCH (Illus. 4-1–4-5) is based on the traditional Scandinavian workbench. It has been a favorite of cabinetmakers for hundreds of years. If you plan on using your bench exclusively for fine cabinetmaking, this is the best design to use. This is a bench that you can be proud of and use for years. As your skills develop, you will appreciate the time-tested design even more.

The shoulder and end vises use commercially available hardware. Buy the hardware before you start building the bench. You may have to alter some of the dimensions given in the plans if the hardware you buy differs from the hardware I used.

Illus. 4-4 shows how to make a tail vise using wooden guides. I also explain how to use commercial steel guides in the Optional Construction Methods section on pages 59–64.

The dog holes in this bench are designed to fit ¾-inch round dogs. If you would rather use traditional square dogs, follow the directions in the Optional Construction Methods section.

I have simplified some of the joinery used for this type of bench, but none of the changes affect the basic design of the bench. Traditionally, this type of bench would use dovetails or finger joints for the vises and mortise-and-tenon joints on the base. However, I found

Illus. 4-1. This classic cabinet-maker's workbench is based on a traditional Scandinavian design.

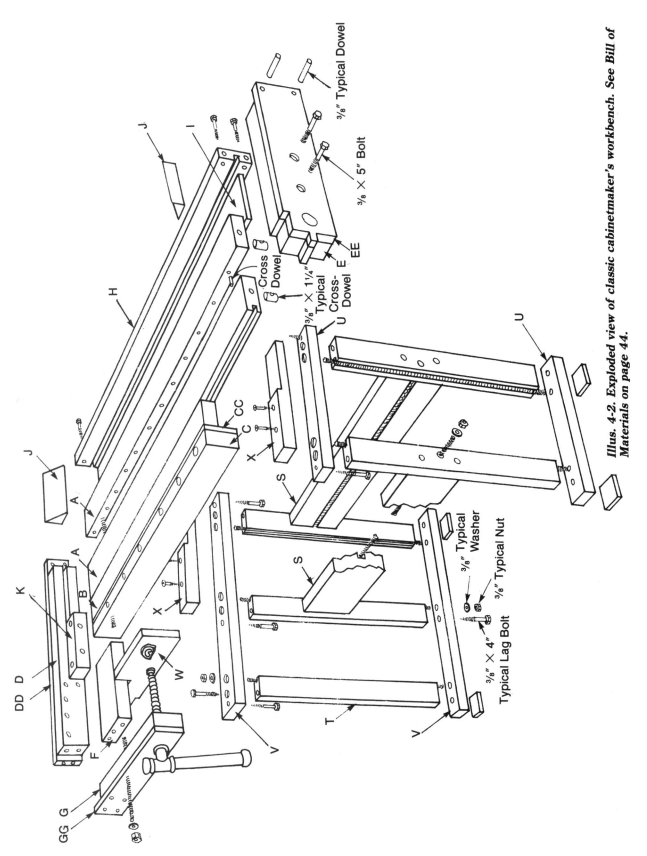

Illus. 4-2. *Exploded view of classic cabinetmaker's workbench. See Bill of Materials on page 44.*

3/8" Typical Dowel

3/8 × 5" Bolt

3/8" × 1¼" Typical Cross-Dowel

Cross Dowel

3/8" Typical Washer

3/8" Typical Nut

3/8" × 4" Typical Lag Bolt

42

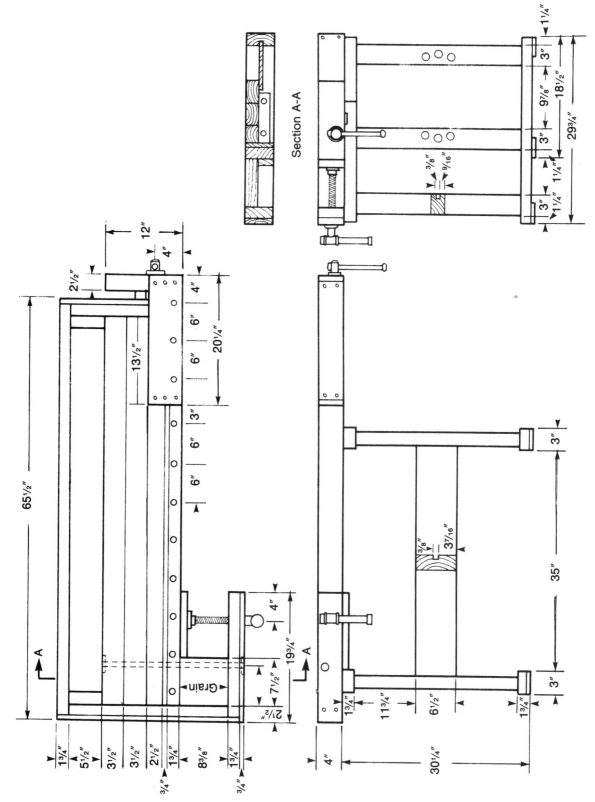

Section A-A

Illus. 4-3. Plans for the classic cabinetmaker's workbench.

43

Classic Cabinetmaker's Bench Bill of Materials

	Description	Size (Inches)	Material	Number Required
A	Bench-Top Core	1¾ × 3½ × 60	8/4 Maple	2
B	Bench-Top Core	1¾ × 2½ × 46½	8/4 Maple	1
C	Dog-Hole Strip	1¾ × 4 × 46½	8/4 Maple	1
CC	Dog-Hole-Strip Cap	¾ × 4 × 46½	4/4 Maple	1
D	Left Frame Member	1¾ × 4 × 26	8/4 Maple	1
DD	Left Frame Cap	¾ × 4 × 29⅜	4/4 Maple	1
E	Right Frame Member	1¾ × 4 × 15½	8/4 Maple	1
EE	Right Frame Cap	¾ × 4 × 17¼	4/4 Maple	1
F	Support Block	1¾ × 7½ × 8⅜	8/4 Maple	1
G	Shoulder Arm	1¾ × 4 × 19	8/4 Maple	1
GG	Arm Cap	¾ × 4 × 19¾	4/4 Maple	1
H	Rear Frame Member	1¾ × 4 × 64	8/4 Maple	1
I	Tool Tray	½ × 7 × 61	3/4 Maple	1
J	Dust Ramp	1¾ × 1¾ × 5½	8/4 Maple	2
K	Left End Cleat	1¾ × 1¾ × 8	8/4 Maple	1
L	Tail Vise Arm	1¾ × 4 × 10¼	8/4 Maple	1
LL	Tail Vise Arm Cap	¾ × 4 × 12	4/4 Maple	1
M	Tail Vise Jaw	1¾ × 3½ × 8¾	8/4 Maple	1
MM	Tail Vise Jaw Cap	¾ × 4 × 10½	4/4 Maple	1
N	Tail Vise Dog Strip	1¾ × 3½ × 18¾	8/4 Maple	1
NN	Guide Runner	¾ × 3½ × 15¼	4/4 Maple	1
O	Tail Vise Top	½ × 5 × 18¾	3/4 Maple	1
P	Lower Guide	¾ × 1¾ × 17¾	4/4 Maple	1
Q	Lower Guide Support	1¾ × 2½ × 5¼	8/4 Maple	1
R	Upper Guide	¾ × 1½ × 15	4/4 Maple	1
S	Stretcher	1¾ × 6½ × 36	8/4 Maple	2
T	Leg	1¾ × 3 × 26¼	8/4 Maple	5
U	Right Rail	1¾ × 3 × 18½	8/4 Maple	2
V	Left Rail	1¾ × 3 × 29¾	8/4 Maple	2
W	Shoulder Vise Jaw	¾ × 5 × 17	4/4 Maple	1
X	Filler	1¾ × 2¼ × 15	8/4 Maple	2

Classic Cabinetmaker's Bench Hardware List

Number Required	Description
1	Shoulder Vise Screw (11″ Yost or Similar Screw)
1	Tall Vise Screw (17″ Record or Similar Screw)
6	$3/8$″ Threaded Rod (6-Foot Lengths)
2	$3/8$″ Dowel (4-Foot Lengths)
22	$3/8 \times 4$″ Lag Bolts
6	$1/4 \times 3$″ Lag Bolts
2	$3/8 \times 5$″ Bolt and Cross-Dowel Fasteners
16	$3/8$″ Nuts
39	$3/8$″ Washers
6	$1/4$″ Washers
1	$1/2$″ Washer
32	#8 $\times$ $11/4$″ Wood Screws
4	#12 $\times$ $21/4$″ Wood Screws
4	#12 $\times$ 3″ Wood Screws

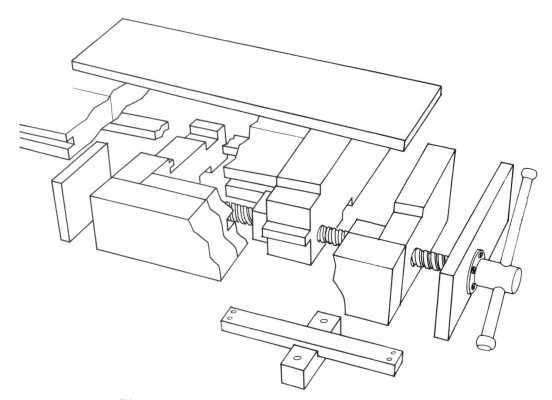

Illus. 4-4. Details of classic cabinetmaker's workbench.

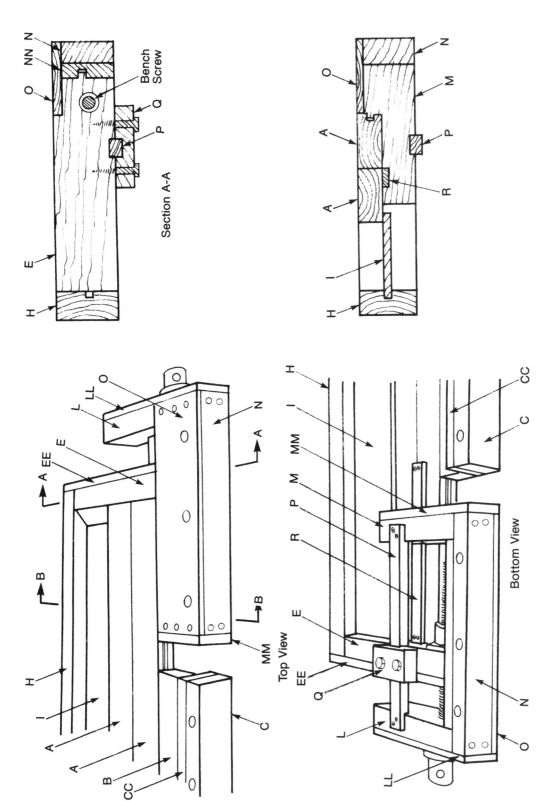

Section A-A

Bench Screw

Top View

Bottom View

Illus. 4-5. Details of classic cabinetmaker's workbench.

46

a drawing of an 18th-century bench that used peg joints in a tail vise, so I decided to test them. They proved to be strong enough, so I included peg joints in the design of this bench.

The base uses truss-rod-reinforced butt joints. Truss rods are one of the best ways to reinforce a workbench base. If you are a more advanced woodworker and would like to use the traditional joints, I explain the necessary changes in the Optional Construction Methods section on pages 59–64.

The frame members are 2½ inches thick. Lumber this thick is hard to find and when you do find it, it is often not completely dry, so I recommend that you glue up these pieces from thinner boards. Gluing up the frame members also simplifies the joinery, because you can make the rabbets by using a longer board on the outside face.

BUILDING THE BENCH TOP

This bench top is made of solid wood, so it is more difficult to make than the one described in Chapter Three, but it is not beyond the skill of most beginning woodworkers.

Start by making the core. The core (Part A) is made from two 1¾-inch-thick maple boards. After you have cut the boards to size, use a router or a plow plane to cut a ½-inch-wide, ⅜-inch-deep groove in the front edge of the front board for the tail vise (Illus. 4-6). Refer to Illus. 4-2 for the exact location of the groove. It is much easier to make this groove now than to try to cut it after the bench top is assembled.

Now, prepare the boards for gluing. As shown in Illus. 4-2, place the two Part A board and the Part B boards next to each other; then examine their end grain. To minimize warping, position the boards so that the curve of the rings in the end grain alternates in direction from one board to the next (Illus. 4-7). Make a mark on the top face of each board so that you can assemble the boards with the proper grain orientation.

Keeping the boards properly aligned as you glue them up will minimize the amount of planing necessary to flatten the top later, so you should use dowels spaced 6 inches apart to keep the boards aligned as you clamp them. Use a dowelling jig to guide the drill; place the fence of the jig against the top face of the boards (Illus. 4-8). That way, the top faces will be flush even if the boards are of slightly different thicknesses.

Before assembling the boards, drill a ½-inch hole for the shoulder-vise truss rod (Illus. 4-9). This hole goes completely through the bench top; when the rod is threaded through it, it will fit the shoulder-vise arm in place. It is difficult to drill this hole after the top is assembled, so drill through each board before you assemble the top. Carefully mark the location of the hole on the edge of each board. All of the holes must line up after assembly.

Illus. 4-6. Before gluing up the core, cut a groove in the front edge for the tail vise.

Illus. 4-7. To minimize warping, alternate the direction of the growth rings in the boards.

Illus. 4-8. Dowels help keep the boards in alignment as you glue up the bench top. Use a dowelling jig to guide the drill.

Illus. 4-9. Before assembling the boards, drill the hole for the shoulder-vise truss rod.

Apply a coat of aliphatic resin glue to both mating edges and insert the dowels. Place the boards flat on a pair of sawhorses and use bar clamps to pull the joint tight (Illus. 4-10).

Illus. 4-10. Apply glue to the mating surfaces, and then use bar clamps to pull the joints tight.

After the glue is dry, remove the clamps and follow the same procedure to attach Part B. Be sure to drill the shoulder-vise truss-rod hole before assembly.

Next, glue up the dog-hole strip. It is made from one 1¾-inch-thick board (Part C) and one ¾-inch-thick board (Part CC). The 1¾-inch-thick board forms the front edge of the bench top. Clamp the boards together until the glue dries.

When the glue is dry, remove the clamps and drill the hole for the shoulder-vise truss rod. Now, attach the dog-hole strip to the bench top core. Drill dowel holes spaced 6 inches apart to align the parts. Make sure that the dowels won't interfere with the dog holes you will drill later. You can use a standard dowel jig to drill the holes in the edge of Part B, but the dog-hole strip is too wide for the jig. You must lay out the location of each hole and drill it without the jig. You can use a drill guide to keep the drill square with the face (Illus. 4-11). When you have all of the dowel holes drilled,

Illus. 4-11. Next, attach the dog-hole strip to the bench-top core. Drill dowel holes to align the parts.

spread glue on the mating surfaces and assemble the parts. Use bar clamps to pull the joints tight.

Now, add the shoulder-vise support block (Part F) to the bench-top core. Notice how the grain is oriented in Illus. 4-3. The grain in the support block runs at a right angle to the grain in the bench top. Orienting the grain in this way minimizes the effects of shrinkage and swelling in the bench top. Wood shrinks and swells mostly across the grain. The dimension with the grain doesn't change very much. If you ran the grain in the support block parallel to the grain in the bench top, the dimensional change in that section of the top would be equal to the change in a board about 20 inches wide. By running the grain at a 90-degree angle, you ensure that the dimensional change in the bench top is limited to a 12-inch-wide area.

Drill two dowel holes and the truss-rod hole next. The truss-rod hole is too deep for most drill bits, so you must drill from both ends. Insert the dowels and the threaded rod in the bench top. Then apply glue and put the support block in place. Put a washer and nut on the back of the rod and push them into the countersunk hole. Make sure that you leave enough rod extending from the front of the block so you can attach the vise arm later.

Now, clamp the block in place. You can temporarily tighten a nut on the front of the truss rod to help clamp the support block as the glue dries. You have to remove the nut before attaching the shoulder-vise arm.

The ends of the bench-top core must be straight and square. If necessary, trim them with a portable circular saw. Use a straight board clamped to the core as a guide. Measure the distance from the edge of the

saw base to the blade and offset the guide this distance. Press the base of the saw against the guide board as you make the cut.

BENCH-TOP FRAME

The bench-top frame supports the bench-top core. It also forms part of the shoulder-vise and tail-vise guides. The width of a solid-wood core will change significantly due to changes in humidity. The frame is designed to allow the core to shrink and swell without affecting the outside dimensions of the bench top.

The ends and the shoulder-vise arm are glued up from two boards. You can simplify the joinery by gluing up the boards while assembling them on the bench top. However, if you choose to assemble them this way, clamp the mating boards together as you drill the holes for the bench screws and cut the tail's vise guide so that they will be aligned when you assemble them.

After you have cut the parts to size, cut the grooves for the tool tray. You can use a router, table saw, or plow plane (Illus. 4-12) to cut the groove. To cut the stopped groove in the ends with a plow plane, you must cut a small section at the end of the groove with a chisel before you begin planing.

Clamp Parts G and GG together temporarily while you drill the hole for the shoulder-vise screw. The bench-screw nut attaches to the inside face of the shoulder-vise arm with four screws. You can mount the nut on the surface or make a mortise so that the nut is flush with the face of the board. Chop the mortise

Illus. 4-12. The tool tray fits in a groove in the bench-top frame. You can use a plow plane, router, or table saw to cut the groove.

and drill the pilot holes for the screws now, because it will be difficult to do this after assembly.

Clamp Parts E and EE together temporarily as you drill the hole for the tail-vise bench screw and cut the tail-vise guide. A band saw is probably the best tool to use to cut the guide, but a handsaw can also be used.

The frame joints are reinforced by pegging. A pegged joint is a dowel joint that extends through the face of the board. The end of the dowel shows, but that doesn't matter on a workbench. A pegged joint is stronger than a blind dowel joint and is easier to make because you can drill the holes after the parts are assembled and clamped (Illus. 4-13).

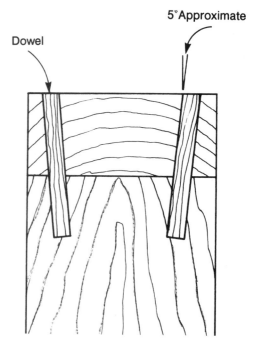

Illus. 4-14. Angling the dowel holes creates a dovetail effect that strengthens the joint.

Illus. 4-13. A peg joint is easy to make, because you can drill the dowel holes after the parts have been assembled.

For added strength, the dowel holes should be angled slightly toward the middle of the board. This creates a dovetail effect that makes it difficult to pull the boards apart after the dowels have been inserted (Illus. 4-14).

Once you have drilled the holes, put a small amount of glue in the joint and use a sliver of wood to spread the glue around in the hole. Don't use too much glue, because it will fill up the bottom of the hole and make it difficult to fully insert the dowel. However, you should make sure that the sides of the hole are completely coated with glue.

Cut the dowel slightly longer than necessary, and then drive it into the hole with a hammer (Illus. 4-15). After the glue is dry, trim the dowel with a handsaw and sand it flush with the surface using a belt sander.

Begin attaching the frame to the core with the left-

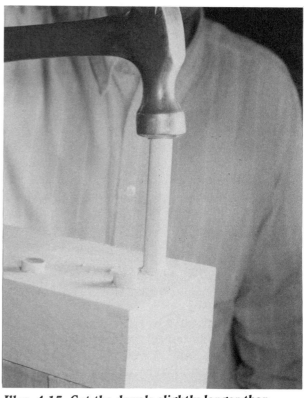

Illus. 4-15. Cut the dowels slightly longer than necessary, and then drive them into the holes with a hammer. After the glue is dry, you can trim the dowel flush with the surface.

end frame member (Part D). Since the grain in the shoulder-vise support block runs parallel to the frame member, you can glue them together without fearing that cross-grain dimensional changes will cause splitting later. However, you should not use any glue on the section that attaches to the end grain of the bench-top core. If this section is not free to move, the top may split later as the boards dry out.

Line up the end of the frame member with the front of the support block and clamp the boards together. Now, drill 3/8-inch-diameter dowel holes through the frame member into the edge of the support block and the end of the dog-hole strip. Spread glue inside the holes and drive in the pegs.

Attach the frame member to the bench-top core with a cleat (Part K). Notice that Illus. 4-2 shows that the rear hole is enlarged to form a slot. This allows the lag bolt to slide back and forth as the top shrinks and swells. To make the slot, drill two holes side by side 1/2 inch apart, and then clean out the wood between the holes with a chisel (Illus. 4-16). Don't use any glue on the cleat.

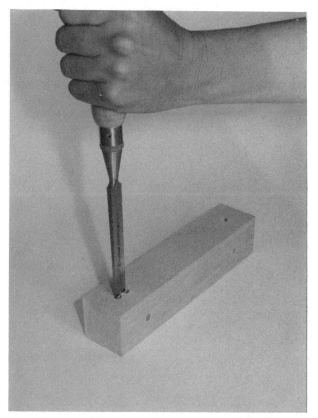

Illus. 4-16. A slotted hole in the cleat allows for dimensional change in the bench top. To make the slot, drill two side-by-side holes and clean out the wood between the holes with a chisel.

Use 1/4-inch lag bolts to attach the cleat (Illus. 4-17). All of the bolts need a 1/4-inch washer, but the bolt in the slotted hole should also have a 3/8-inch and a 1/2-inch washer. The larger washers prevent the 1/4-inch washer from digging into the wood. If the washer were to dig in, it would hold the lag stationary and the advantage of the slotted hole would be negated.

Illus. 4-17. Attach the cleat with lag bolts. Use a large washer to keep the bolt head from digging into the slotted hole.

Next, attach the shoulder-vise arm. Make sure that the end of the left frame member is flush with the front of the support block. If necessary, trim it flush. Drill a 1/2-inch-diameter hole in the arm for the truss rod. Remove the nut and washer that you placed on the truss rod when you attached the support block. Apply glue to the mating surfaces and position the arm. Replace the washer and nut and tighten the nut with a socket wrench (Illus. 4-18). Use bar clamps to apply uniform pressure to the joint.

Drill 3/8-inch-diameter holes and drive two pegs into the support block now, but don't drill the peg holes in the corner joint until after the arm cap has been glued in place.

Next, attach the right frame member. Don't use any glue on this frame member. It will receive all of the thrust from the tail vise, so it must be attached to the core more firmly than the rest of the frame members. I use bolts and cross dowels to give this joint added strength. You can buy cross dowels from a well-stocked woodworking store or from a mail-order woodworking supply company. You can also make your own cross dowels. To do this, you will need a piece of 5/8-inch steel

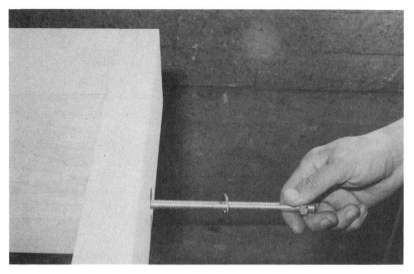

rod, a drill press, a ⁵⁄₁₆-inch drill bit and a ³⁄₈-inch, 16 NC tap. It is easier to drill and tap the holes before you cut the rod into small sections. Lay out the locations of the holes and drill them on the drill press (Illus. 4-19).

Next, tap the holes. Apply cutting oil to the tap and advance it into the hole. Whenever you feel increased turning resistance, back off one-fourth turn to break the chip, and then continue cutting. After you have cut all of the threads, use a hacksaw to cut the cross dowels to length.

Drill the ¹³⁄₃₂-inch bolt holes in the frame member first. Notice from referring to Illus. 4-2 that the rear hole is slotted. Make the slot by drilling two holes ½ inch apart and cutting the wood out between the holes with a chisel. This allows the core to shrink and swell without splitting. Temporarily clamp the end cap in

place and drill the ³⁄₈-inch holes into the core. When drilling the rear hole, position the bit in the middle of the slot.

Now, remove the left frame member and lay out the position of the cross-dowel holes. Use a square to transfer the centerline of the holes in the edge to the underside of the core. Then measure in from the end 1¾ inches. The holes are ⁵⁄₈ inch in diameter and 1¼ inches deep.

Drill the holes for the cross dowels using a Forstner bit or another type of bit that doesn't have a long center spur, so that you won't drill through the top of the core. Now, attach the right frame member using the bolts and cross dowels (Illus. 4-20). Place a large washer on the rear bolt so that it can slide freely in the slot.

Next, slide the tool tray into the grooves in the frame

Illus. 4-19. You can make cross dowels from a steel rod. Drill the holes with a drill press and thread them with a tap.

Illus. 4-20. Use bolt-and-cross-dowel fasteners to attach the right frame member.

members. The tool tray can be made of either ½-inch plywood or ½-inch maple lumber. In either case, don't use any glue in the grooves. The tool tray will be attached to the bench-top core, and it must be free to slide in the grooves as the bench top shrinks and swells.

Attach the tray to the core with 1¼-inch #8 wood screws spaced 6 inches apart. Make sure that there is ¼ inch clearance between the edge of the tool tray and the end of the cleat on the left frame member. If the tool tray hits the end of the cleat as the top shrinks, the top may split. Now, drill pilot holes for the screws and drive in the screws. The screws can be driven in easier if you rub some beeswax on the threads first.

The dust ramp (Part J) should not be glued in place. In this design, the tool tray must slide freely in the groove to prevent the top from splitting. Secure the dust ramp with one screw in its middle. Be sure to leave clearance between the end of the dust ramp and the rear frame member in case the core swells.

Now, attach the rear frame member. Don't use any glue on this joint; if glue is accidently squeezed into the tool-tray groove, it could cause the top to split later. Instead of pegging the rear joints, secure them with ⅜-inch lag bolts. First, drill a 1-inch-diameter hole ½-inch deep to countersink the bolt heads, and then drill the ⅜-inch shank clearance holes through the rear frame member.

Drill ²¹⁄₆₄-inch-diameter pilot holes into the ends of Parts E and D after the rear frame member is in place. Use a washer on each lag bolt and rub wax on the threads before you drive them. If necessary, trim the ends of the rear frame member flush.

Now, prepare the end caps (Parts EE and DD) for attachment. Drill 1-inch holes in the right cap to clear the bolt heads. Enlarge the rear hole on the right end

cap so that the washer can slide when the bolt moves in the slotted hole. Now, apply glue to the mating surfaces and clamp the end caps in place. Drill the peg hole 1¼ inches deep. Notice from referring to Illus. 4-21 how these pegs lock the joint. Pegs are best at resisting shear forces. By driving pegs in both directions, you ensure that one set of pegs will always be resisting a shear force.

To complete the frame, install the shoulder-vise-arm cap (Part GG). Make sure that the end of Part DD is flush with the front of the shoulder-vise arm. Trim it flush if necessary.

Now, apply glue to the mating surfaces and position the cap. Align the hole for the vise screw and the clearance hole for the truss-rod nut with the holes in the arm and clamp the cap in place. Drill the peg holes at the corner. Notice by referring to Illus. 4-3 that these pegs are 4 inches long. They extend through the cap and arm into the end of Part D.

BUILDING A TAIL VISE

In this section I describe how to build a tail vise with wooden guides. If you elect to use the optional steel guides, skip this section and refer to the Optional Construction Methods section on pages 59–64.

Begin by cutting a groove in the guide runner (Part NN) and attaching it to the inside of the dog-hole strip (Part N). Glue the runner in place. You can reinforce the joint with screws, if you position them so they won't interfere with the dog holes that you will drill later.

Next, lay out the guides and cut them to shape. Accuracy at this stage is important if you want a smooth-working vise. To lay out the position of the

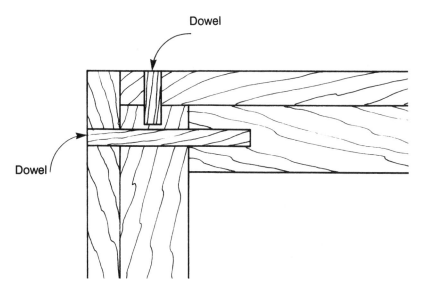

Dowel

Dowel

Illus. 4-21. Driving pegs in two directions locks the joint.

guide tongues, hold the vise parts in place on the bench and mark the location of the guides (Illus. 4-22). Cut the guide tongues with a band saw or a handsaw. Then trim and smooth them with a chisel (Illus. 4-23).

Apply glue to the mating parts and clamp them together. Drill the peg holes and drive in the pegs (Illus. 4-24). Trim all of the joints flush. Then glue on the top and the end caps.

Now, install the bench screw and thread the nut on. Tighten the screw until the vise is fully closed and then mark the location of the screw holes in the nut. Remove the vise and secure the nut to Part E with screws.

Now, reinstall the vise and slide the upper guide strip (Part R) in place. Attach the guide with screws. Don't use any glue, because you must be able to remove this guide to remove the vise. Test the vise to make sure that it still operates smoothly. Make any neces-

sary adjustment. Then close the vise and lay out the position of the lower guide strip (Part P).

You will have to make a cutout in both vise ends and the right frame member for the lower guide. Use a backsaw to make the side cuts. You can cut all the way to the bottom in the frame member, but you have to angle the cut in the vise ends because the cut doesn't continue entirely to the other edge. Now, use a chisel to remove the wood between the saw kerfs (Illus. 4-25). Smooth the bottom by taking fine cuts with the chisel.

Test the fit of the guide strip. Trim the cutout until the guide fits snugly but doesn't bind. Attach the guide to the vise with screws. Don't use glue, because you must be able to remove this guide to remove the vise. Make a corresponding cutout in the lower guide support (Part Q) and attach it with ¼-inch lag bolts (Illus. 4-26).

Illus. 4-22. To ensure an accurate fit, lay out the position of the guide tongues with the other vise parts in place.

Illus. 4-23. After you cut the guide tongues, use a chisel to trim them to fit into the groove snugly.

Illus. 4-24. Assemble the vise with dowels.

Illus. 4-25. Use a chisel to chop out the waste in the cutout for the lower guide strip.

Illus. 4-26. Attach the lower guide support with lag bolts.

Now, test the operation of the vise. You may have to remove it and trim some of the guides to get it operating smoothly. Once you are satisfied with the operation of the vise, lock the bench screw in place by installing the garter. The garter is a split collar that attaches to the vise with screws. It fits in a groove in the bench screw.

DOG HOLES

I used ¾-inch round dog holes. I like using round dog holes because you can attach many different types of hold-downs and bench accessories besides the standard dogs to them. (See Chapter 5.) If you would rather use the traditional square dogs, skip this section and refer to the Optional Construction Methods section on pages 59–64.

Lay out the dog-hole locations as shown in Illus. 4-3. I chose a 6-inch spacing, but you can space them closer if you like. If you change the dog-hole spacing, you must plan for it before beginning construction, so you can position the dowels and screws that attach the dog-hole strip so that they are out of the way.

The dog holes must be drilled accurately and square with the bench top. Use a drill guide to keep the drill straight (Illus. 4-27).

Illus. 4-27. After the tail vise is fully assembled, drill the dog holes using a drill guide.

Drill the holes from the top until the center spur of the bit breaks through the bottom. After you have drilled all of the holes this far, turn the bench top over and finish drilling the holes by drilling into the bottom.

BUILDING THE BASE

The base for the workbench described and illustrated in this chapter is made of maple, but any hardwood will work. The legs, sled feet, top rails, and stretchers are made from 1¾-inch-thick hardwood. The exact thickness is not critical, so if you can get boards that are slightly thicker, use them. You don't have to plane them down to 1¾ inches.

Traditionally, the base would be assembled with wedged mortise-and-tenon joints. In this simplified design, butt joints reinforced with truss rods are used. Truss rods make the base very strong. They are threaded steel rods that tie the joints together (Illus. 4-28). They act like the reinforcing bars in reinforced concrete. Wood is much stronger when compressed than when under tension. Most traditional joints rely on the wood to resist tensional forces. In a truss-rod

Illus. 4-28. Truss rods are threaded steel rods that tie the joints together.

Illus. 4-29. Use a one-inch bit to drill the countersinking holes.

joint, the truss rod prestresses the joint in compression. All of the tensional forces are withstood by the steel, and the compression forces by the wood. The lag bolts in the joint are simply to keep the parts in alignment. Another advantage of truss rod joints is that they can be assembled and disassembled repeatedly, so you can take the bench apart to move it.

For this project, I used ⅜-inch-diameter threaded rod. You can buy this at most hardware stores. The rod comes in 3- and 6-foot lengths. You will have to buy two of the six-foot lengths, because the rod for the stretchers is longer than three feet. A six-foot rod can be cut into one stretcher rod and one leg rod.

Cut the parts to the lengths shown in the Bill of Materials first, and then rip them to their widths. Smooth the cut edges with a hand plane or a belt sander. Now, cut the grooves for the truss rods. You can use a router with a ⅜-inch straight bit, a dado blade on the table saw, or a plow plane. See Illus. 4-2 for the location of each groove.

Now, mark the location of all the truss-rod and lag-bolt holes. It is important to drill these holes accurately. If you have a drill press, use it to drill all the truss-rod holes. If you don't have a drill press, use a drill guide mounted on a portable drill. Drill the countersinking holes first (Illus. 4-29). Use a 1-inch bit and drill into the wood ½ inch. Next, drill a ¹³/₃₂-inch hole through the board. The holes are oversize, to allow for wood shrinkage later.

Now, you are ready to assemble the base. Begin with the legs. Put the threaded rod in the groove in the legs. Then insert the rods into the holes in the top rail and sled foot. Place a flat washer on one end of

the threaded rod and screw on the nut (Illus. 4-30). Fit the nut snugly against the washers, and then use a socket wrench to tighten the nut.

Illus. 4-30. Place a flat washer and nut on the end of the truss rod.

Next, drill $21/64$-inch pilot holes for the lag bolts (Illus. 4-31). Place a washer on the bolt and rub the threads with beeswax to make the bolt easier to drive into the hardwood. Put the lag bolts in the pilot holes and tighten them with a socket wrench (Illus. 4-32). Next use a hacksaw to cut off the excess truss rod. Follow the same procedure to attach the stretchers.

Illus. 4-31. Lag bolts prevent the joints from twisting out of alignment. Drill the pilot holes after you have tightened the truss rods.

Illus. 4-32. Insert the lag bolts in the pilot holes and tighten them with a socket wrench.

After the base is assembled, attach the feet. Temporarily clamp them in place as you drill pilot holes for the screws. Countersink the holes so that the screw heads will be below the surface, and then drive the screws (Illus. 4-33).

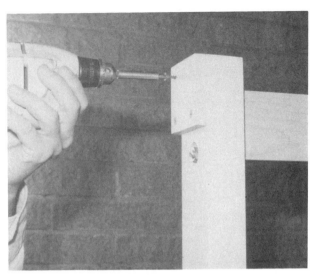

Illus. 4-33. Attach the feet with screws.

At this point, place the bench top on the base. Attach the spacer blocks to the top of the rails. The bench top varies in thickness. The spacer blocks support the area in the middle of the bench top. Since the top is solid wood, it can change dimensions slightly as its moisture content changes. If you don't allow for this movement, it can cause the top to split or warp. For this reason, the core "floats" in the frame. Don't use any glue or drive any lag bolts into the bench-top core. All of the lag bolts should be driven into the bench-top frame. This will leave the core free to shrink or swell and allow you to firmly attach the top to the base.

PLANING THE BENCH TOP

When you make a bench top out of solid wood, you must plane it flat after the bench is assembled. If you were careful during assembly, you won't have to do a lot of planing.

The first step is to check the bench top with a straightedge. Place a long straightedge on the bench top and look for high spots. (High spots are areas of the bench top that are thicker or wider than the rest of the bench top.) Make a pencil mark on the high spots and move the straightedge to another location.

Next, check for winding. You will need two boards about 2 inches wide and 36 inches long. The edges of the boards should be straight and square. These boards are called *winding sticks*. Place one at each end of the bench and then sight across them. If the edges of the winding sticks are parallel, then the top is flat. If the winding sticks are not parallel, mark the high corners on the bench top. You will have to plane the high corners to flatten the bench top.

If the top is flat and you only have to remove a small amount of wood, use a belt sander. If you must remove a lot of wood, use a good hand plane. A long plane like a try or a jointer plane works best, but you can also do the job with a jack plane (Illus. 4-34). Start by cutting down the high spots you marked. Next, take long strokes the full length of the bench top. Check your progress frequently with a straightedge and winding sticks. When the top is flat, adjust the plane for a very fine cut and smooth the surface. You can give the top a final smoothing with a cabinet scraper.

Illus. 4-35. The space between the shoulder-vise support block and the top rail of the base acts as a guide for the shoulder-vise jaw.

Illus. 4-34. Use a long plane to flatten the bench top.

SHOULDER VISE

To complete the shoulder vise, make the vise jaw from a piece of ¾-inch-thick maple. Cut its end to fit in the space between the support block and the top rail of the base (Illus. 4-35). Trim or shim it as necessary to get a good fit. Attach the iron socket to the vise jaw and attach the vise jaw to the end of the vise screw. The vise jaw should slide smoothly as you advance the screw. If it binds, adjust the guide and rub wax inside the groove.

Finally, sand the bench smooth and give it a penetrating oil finish to complete the job (Illus. 4-36).

OPTIONAL CONSTRUCTION METHODS

In this section I present some alternate ways to build the bench. If you prefer to use commercial tail-vise

Illus. 4-36. Seal the wood with several coats of penetrating oil.

guides, you can modify the bench plans using the information in this section. Also included are directions for making square-dog holes for the more traditional style of bench dogs.

As I mentioned at the beginning of this chapter, I simplified the construction of the bench so that even a beginning woodworker can build it. If you are a more advanced woodworker, you may want to "dress up" the bench with more complicated joints. This type of joinery is also covered in this section.

Using Commercial Tail-Vise Guides

If you choose to use commercial steel guides for the tail vise, follow the directions and plans in this section instead of the ones given previously. The exact dimensions of the vise parts will depend on the type of vise hardware you use.

Cut the parts for the tail vise as shown in Illus. 4-37. Drill a 1¼-inch diameter hole for the bench screw in the vise end plate and the end of the vise-screw box.

Now, start assembling the tail vise. Begin by assembling the vise core. The core is made from three strips of wood attached to two thick end blocks. This method of making the clearance for the vise screw and the vise screw nut is simpler than trying to carve out a cavity in a solid piece of wood. Use glue to attach the sides to the end blocks. Now, attach the guide plates to the core. Drill holes all the way through the core for the two mounting bolts.

Attach the tail-vise mounting block to the underside of the bench with glue and lag bolts. Then drill the 1¼-inch-diameter × 1-inch deep bolt-clearance hole in the mounting block. This hole provides clearance for the bolt that attaches the bench screw nut to the bench plate.

Now, mount the bench plate to the bench. Position the bench plate as shown in Illus. 4-37 and attach it with two wood screws in the upper holes. Next, attach the vise core to the bench plate. Loosen the guide-plate mounting bolts slightly and slide the guide plates over the bench plate.

Now, tighten the guide-plate bolts. The guide plates should be snug but should not bind on the bench plate. If necessary, shim the lower guide plate until you get a good fit. Screw the vise screw into the vise screw nut and check to see that the vise operates smoothly. Also check that the vise is level with the bench top.

When you are satisfied with the operation of the vise, make the top cap. Measure the distance from the top of the upper guide plate to the top of the bench

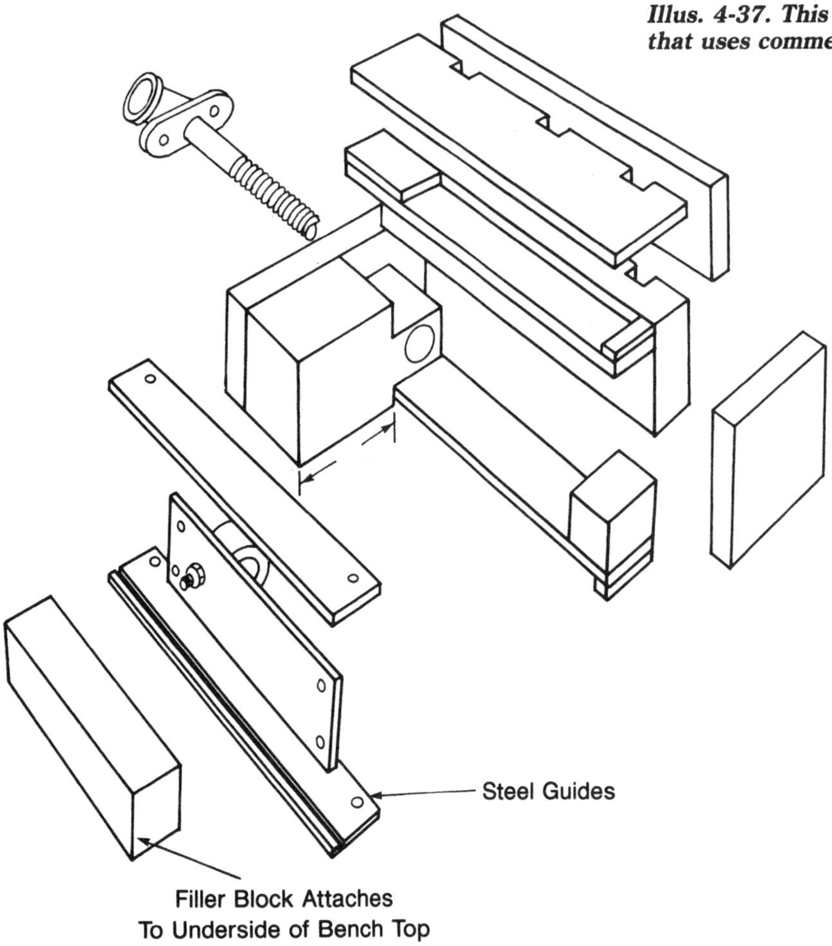

Illus. 4-37. This plan shows how to make a tail vise that uses commercially available steel guides.

Steel Guides

Filler Block Attaches
To Underside of Bench Top

top. This is the thickness of the top cap. Cut filler strips to fill in around the upper guide plate. Leave the vise core attached to the bench as you glue up the rest of the vise. That way, you can make sure that the parts are aligned with the bench.

Use epoxy glue to glue the filler strips and the top cap in place. Coat the top of the steel guide plate with epoxy. Clamp the top cap in place and let the glue cure. After the glue is dry, check to see that the vise is operating smoothly.

Now, prepare to glue the dog-hole strip cap to the front of the vise. Apply glue to the mating surfaces and install the dog-hole strip cap. Use bar clamps across the bench top to clamp the joint.

If you are satisfied with the way the tail vise is aligned with the bench, install the rest of the bench-plate fasteners. Use wood screws in the top row of holes and flathead bolts in the bottom row.

The final step is to install the end caps. Remove the bench screw and put the end cap in place. Now, reinstall the bench screw and tighten it to firmly clamp the end cap as the glue dries. Drill pilot holes and install the four wood screws that secure the end plate. Also install the two wood screws that attach the bench-screw flange to the end cap.

Making Square Dog Holes

Traditionally, this type of bench would have square dog holes. If you want to use square dogs on your bench,

follow the directions below. Make the dog holes before you glue up the dog-hole strips (Parts C, CC, N, and NN in Illus. 4-38 and 4-39). Lay out the dog holes as shown in Illus. 4-40. Notice that the dog holes in the bench slope 2 degrees towards the end vise and the holes in the end vise slope 2 degrees towards the bench. The sloping dog holes help to pull the work tight against the bench as you tighten the vise. Pressure on the dog tends to push it deeper into the hole. You can cut the dog holes by hand or with a table saw or a router.

To cut the dog holes by hand, lay out the position and shape of each hole on the board. Use a backsaw to cut the shoulders, and then remove the waste with a chisel. Use the chisel to enlarge the notch at the top so that the head of the dog will fit flush with the bench top. Test the fit by inserting a dog. Make any necessary adjustments with a chisel, and then move on to the next hole.

To cut the dog holes with a table saw, install a dado blade. Make test cuts and adjust the width of the dado blade until the dog fits snugly in the hole. Adjust the mitre gauge for the 2-degree angle and cut the notches. You will still have to use a hand chisel to enlarge the top of the holes to fit the top of the dog.

You can make a simple jig to guide a router to cut the dog holes (Illus. 4-41). The jig is made from a piece of 1/8-inch-thick hardboard and four strips of 3/4-inch-thick wood. Attach the bottom guides to the hardboard so that they fit snugly against the edges of the dog-hole strip.

Next, attach one of the router base guides. This

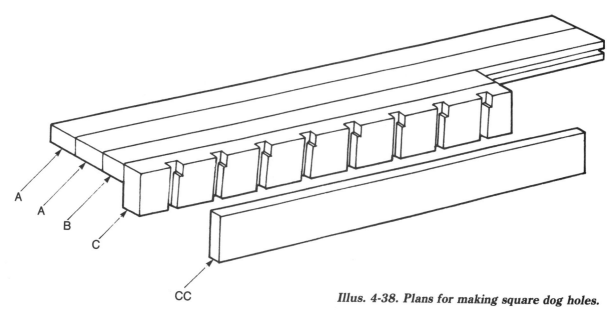

Illus. 4-38. Plans for making square dog holes.

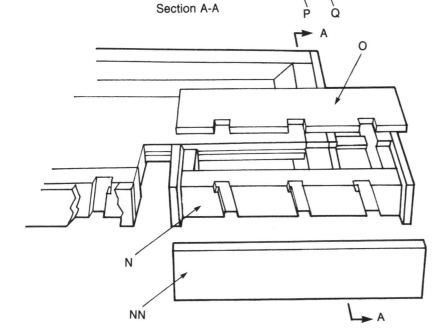

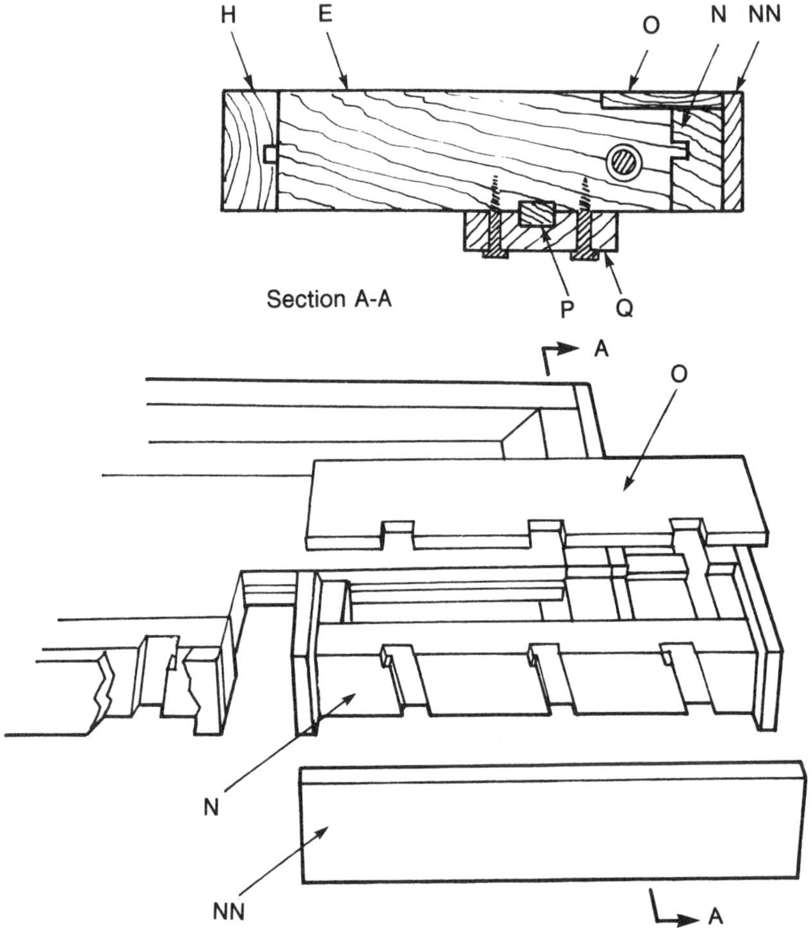

H   E   O   N NN

Illus. 4-39. Plans for making square dog holes.

Section A-A

P   Q

A

O

N

NN

A

2° Slope

1¹⁄₁₆″

1¹⁄₁₆″

1″

Bench

Vise

¹³⁄₁₆″

6″   3″   4″

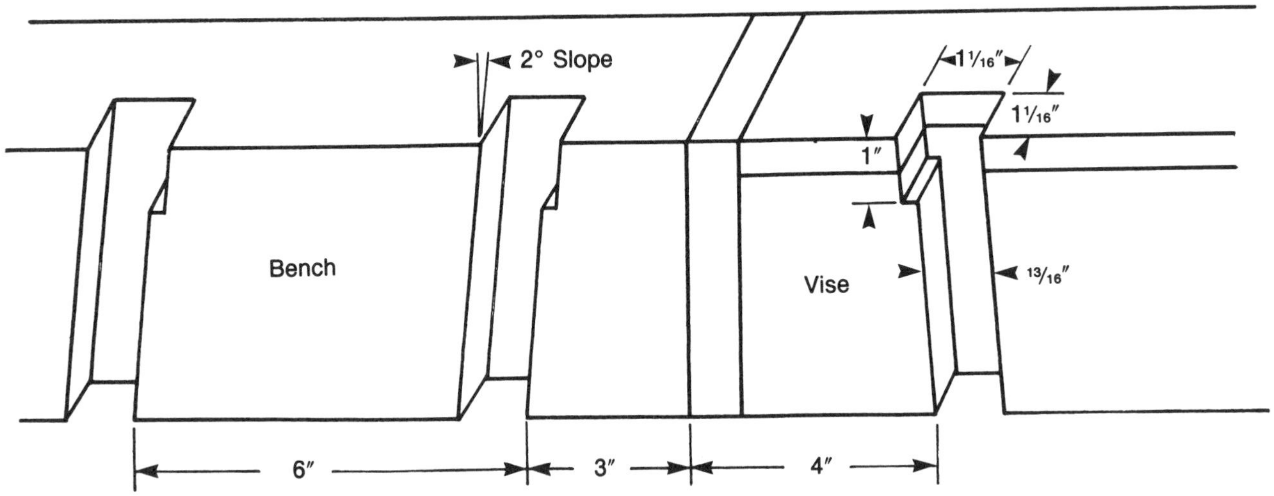

Illus. 4-40. Plans for making square dog holes.

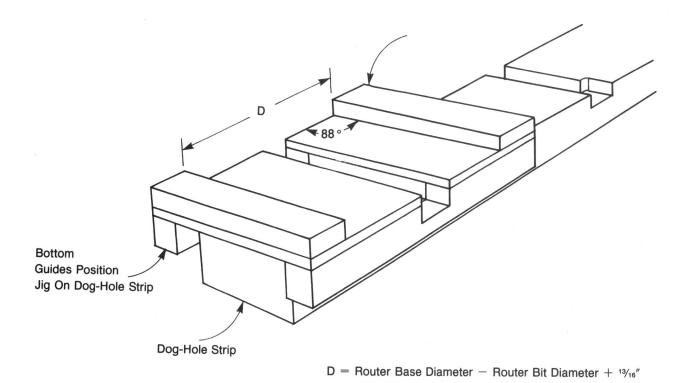

D = Router Base Diameter − Router Bit Diameter + $^{13}/_{16}''$

Illus. 4-41. This router jig guides the router as you cut square dog holes.

guide should be placed 2 degrees off square. Now, place the guide on a scrap board and make a test cut. Use a ³/₈-inch straight bit in the router. The first cut will cut a slot in the base of the jig. Temporarily clamp the other router base guide to the jig. Make test cuts and adjust its position until the dog fits snugly in the cut made by the router. Then secure the guide to the jig. Make sure that it is parallel to the first guide.

To use the jig, lay out the dog-hole locations on the work first, and then place the jig over the first location. Align the slot in the base of the jig over the layout lines, and then clamp it in place. You can cut the notch for the dog hole with the router. Move the jig so that one edge of the slot lines up with the layout line for the dog-hole notch. Clamp the jig in place. With the router off, place it in the slot and align it with the layout line for the bottom of the dog-hole notch.

Next, mark the location of the router base on the jig. Move the router bit away from the wood and start the router. Make the cut and stop when the router base reaches the mark you made on the jig. Cut the rest of the dog holes following the same procedure.

You can't use the same jig to cut the dog holes in the tail vise, because the jig slopes in the wrong direction. Make another jig to cut the tail-vise dog holes or cut them by hand.

After all the dog holes have been cut, glue up the front part of the bench top. Cut the dog-hole backing strip and prepare to glue it to the dog-hole strip. Place a dowel between each dog hole to help keep the boards in alignment as you clamp them. Use a dowel jig to locate the holes.

Now, spread aliphatic resin glue over both mating surfaces and clamp the boards together. Use a damp rag to wipe off any glue that squeezes into the dog holes.

ALTERNATE JOINERY

Although the bench shown in Illus. 4-1 is strong and good-looking, advanced woodworkers may want to show off their joinery skills by using more complicated joints. Illus. 4-42 shows some ideas for alternate joints.

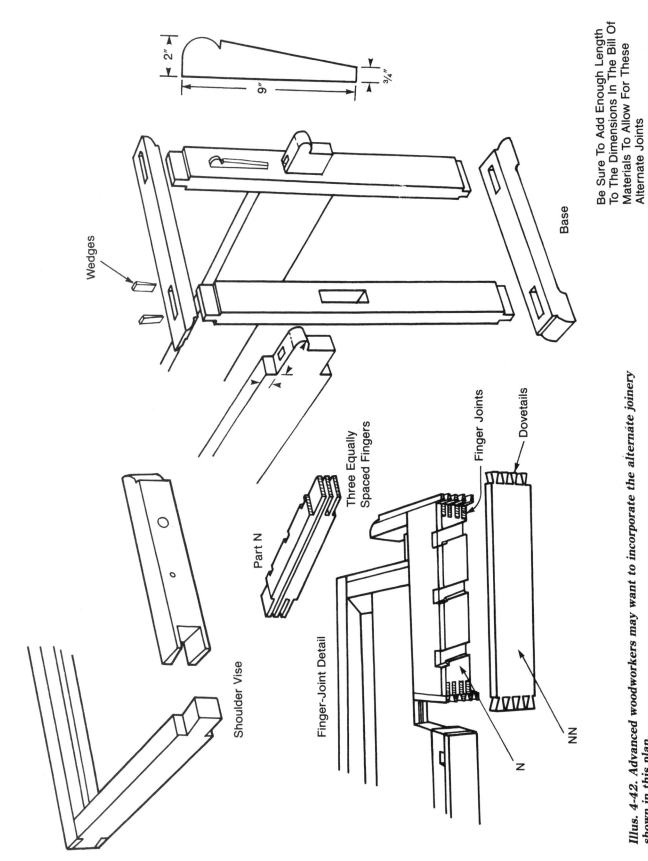

2"

9"

3/4"

Wedges

Base

Be Sure To Add Enough Length
To The Dimensions In The Bill Of
Materials To Allow For These
Alternate Joints

Shoulder Vise

Part N

Three Equally
Spaced Fingers

Finger-Joint Detail

Finger Joints

Dovetails

N

NN

Illus. 4-42. Advanced woodworkers may want to incorporate the alternate joinery shown in this plan.

CHAPTER FIVE

Workbench Accessories

THERE ARE MANY workbench accessories that can make it easier to secure the work to the workbench (Illus. 5-1). Most of these accessories fit into holes in the bench top. In this chapter, I describe some of the most useful accessories and ways to use them.

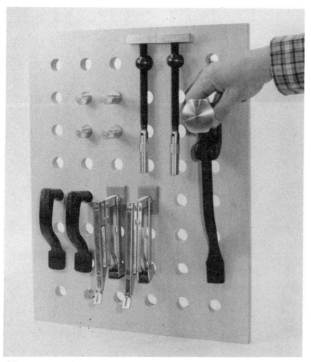

Illus. 5-1. Workbench accessories fit into holes in the bench top. They provide a variety of clamping options. As shown here, you can keep your workbench accessories organized by hanging them on a storage board. To make a storage board like this, drill a grid of ¾-inch holes in a piece of plywood.

BENCH DOGS

You must use bench dogs to clamp a board using the tail vise. There are two basic types of bench dogs: ¾-inch round dogs and square dogs. The square dog is the more traditional one, but the ¾-inch round dog is more versatile, because there are many other bench accessories that will fit in a ¾-inch hole.

Bench dogs can be made from steel, brass, plastic, or wood (Illus. 5-2). The steel and brass dogs are very strong, but if you accidently hit one with the edge of a tool, you can nick the tool's cutting edge. Wood and plastic dogs are strong enough for most work, and they won't mar the wood or damage the cutting edge of your tools.

It's a good idea to have both metal and wood dogs for your bench. Metal dogs grip better, so you can use them when it is important to get the firmest grip possible. When you don't need as tight a grip and don't want to leave a dog mark on the end of the board, use wood dogs.

You can buy ready-made bench dogs, but the wood ones are easy to make. To make round dogs, start with a ¾-inch-diameter dowel and cut it to length. Cut a notch in one end as shown in Illus. 5-3. Notice that there is a 2-degree slope on the face of the dog. This helps to prevent the work from lifting off the bench as you tighten the vise. Drill a hole in the side of the dog and insert a brass bullet catch. This spring-loaded catch will hold the dog in position until you tighten the vise.

To make square wooden bench dogs, use the pattern in Illus. 5-3. Cut the dogs from a strong hardwood such as maple. After you cut them out, drill the hole for the bullet catch in their front edges and drive the catch in (Illus. 5-4).

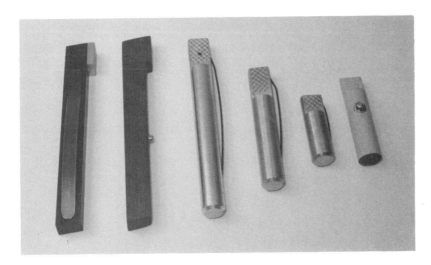

Illus. 5-2. Several different bench dogs. From left to right: a square steel bench dog; a shop-made square wooden bench dog; a ¾-inch round steel bench dog; a ¾-inch round brass bench dog; a short, ¾-inch round brass bench dog; and a shop-made ¾-inch round wooden bench dog.

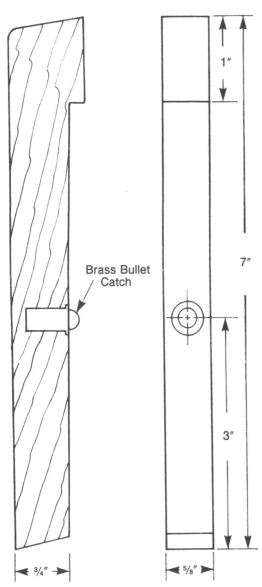

Brass Bullet
Catch

1"

7"

3"

¾"

⅝"

Illus. 5-3 (left). This plan shows how to make your own wooden bench dogs.

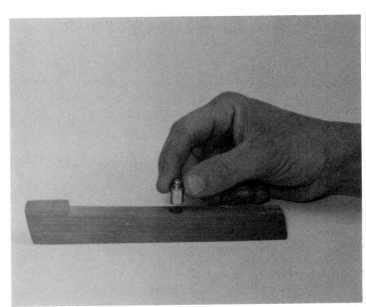

Illus. 5-4. A bullet catch in the edge of a wooden bench dog holds the dog in any position on the bench.

CLAMPING DOGS

Illus. 5-5 shows useful variations of clamping dogs. One such variation is a threaded clamping bar. The ¾-inch-diameter shank of this clamping dog will fit into a standard bench-dog hole. These clamping dogs are sold under the trade name Wonder Dog by Veritas Tool Company. You can use these dogs to clamp between any two holes on the bench top, so they can be used to clamp items with curves and odd shapes. In fact, they can do most of the work of a tail vise, so if you can't afford a tail vise, drill ¾-inch dog holes in your bench top and buy a couple of clamping dogs (Illus. 5-6).

POPPETS

Sometimes you need to work on a part that seems impossible to clamp in a vise or with standard bench dogs. Poppets are a variation on the bench dog that can solve the problem (Illus. 5-7). They fit into standard dog holes, but have center pins that support the work

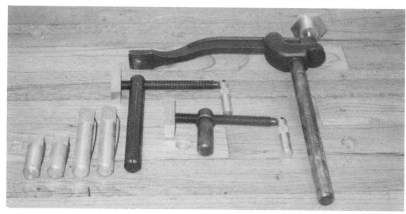

Illus. 5-5. Several workbench accessories made by the Veritas Tool Company. The clamping dogs in the middle are sold under the trade name Wonder Dog.

Illus. 5-6. You can use a clamping dog as a substitute for a tail vise.

Illus. 5-7. You can make a set of poppets like these to hold odd-shaped parts in a tail vise.

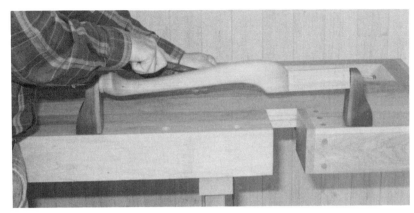

Illus. 5-8. Poppets are handy for holding legs while they are being shaped.

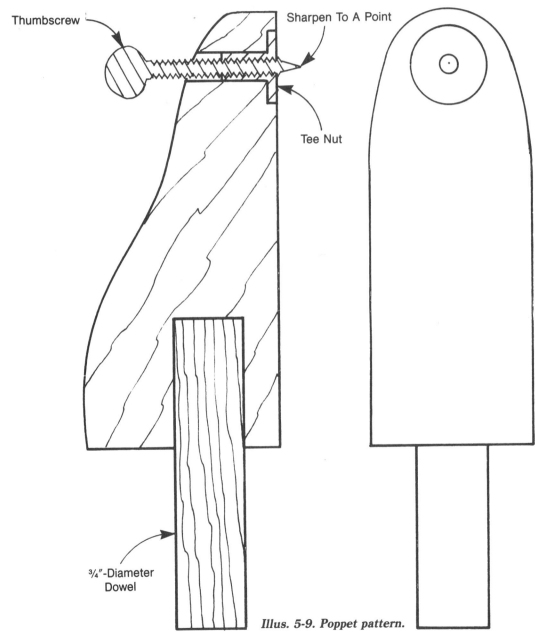

Thumbscrew

Sharpen To A Point

Tee Nut

¾"-Diameter Dowel

Illus. 5-9. Poppet pattern.

above the bench top. They are particularly well suited for carving and shaping legs (Illus. 5-8). Place one poppet in the tail vise and the other in the bench top. Put the part that will be worked on between the steel centers of the poppets and clamp the tail vise tight. You can rotate the work to any position by slightly loosening the tail vise. Tightening the vise will clamp the work securely.

You can easily make a set of poppets. Illus. 5-9 is a full-size pattern. Cut the poppets out of a strong hardwood. Drill a ¾-inch-diameter hole in their bottoms and glue into the hole a piece of ¾-inch-diameter dowel. Drill the holes for the centers and install the Tee nuts. T nuts are threaded inserts that fit into the hole so that you can use a thumbscrew as the center. T nuts can be bought at most hardware stores. Grind the end of the thumbscrews to a sharp point, and then screw them into the T nuts. The point of the screw should extend about ¼ inch past the edge of the poppet to form the center pin.

HOLD-DOWNS

A hold-down can be used to clamp the work to the bench top. The oldest type of hold-down is called a holdfast. André Roubo describes holdfasts in his 1769 book *The Art of the Woodworker*. They are still commercially available today. Commercial holdfasts will work in a ¾-inch dog hole. A holdfast works because its shank is wedged in a hole in the bench top.

Holdfasts are very easy to use and quick to set and release. This makes them very useful for carvers or when you are working on something that needs to be repositioned frequently. To set a holdfast, strike its top with a mallet (Illus. 5-10). To release a holdfast, strike it on its back with the mallet (Illus. 5-11).

Illus. 5-11. To release a holdfast, strike it on its back with a mallet.

Illus. 5-10. To set a holdfast, strike its top with a mallet.

The only drawback to the traditional holdfast is that it tends to wear dog holes, so that the dog holes will not be useful for other bench accessories. It is particularly hard on man-made bench tops, but even dog holes drilled in solid beech will eventually become oversize when you use holdfasts frequently. This isn't a problem if you will only be using the holes for holdfasts, because they will adjust to an oversize hole. However, if you will be using the same holes for bench dogs and other accessories, it is better to use another type of hold-down.

Black & Decker sells the hold-down shown in Illus. 5-12; it will fit in a standard ¾-inch hole. It has a plastic sleeve that protects the hole, so the holes won't wear out. This is a very useful hold-down, and can be easily found at many local tool centers. To operate the hold-down, insert it in a hole and raise the handle. Press the hold-down against the work, and then push down on the handle to lock it.

The Veritas hold-down (Illus. 5-13) looks a lot like a traditional holdfast, but you set it by turning a handle. It is wedged into a ¾-inch dog hole, but it doesn't cause

Illus. 5-12. This Black & Decker hold-down has a plastic sleeve that protects the dog holes.

Illus. 5-14. This Jorgensen hold-down works like a C-clamp. It attaches to a special bolt that is recessed in the bench top.

Illus. 5-13. Turning the handle on this Veritas hold-down adjusts the clamping pressure.

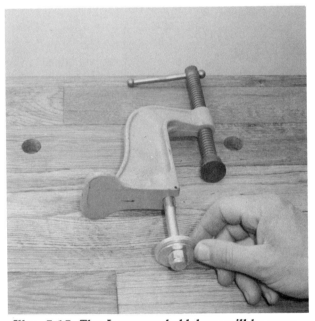

Illus. 5-15. The Jorgensen hold-down will be more versatile if you adapt it to work in ¾-inch dog holes by adding large fender washers.

as much wear as a holdfast. Its long reach and adjustable tension make it very useful.

The Jorgensen hold-down works like a C-clamp (Illus. 5-14). It is designed to attach to a special bolt recessed in the bench top. You can buy extra bolts and place them in several positions on the workbench, but that doesn't provide the versatility that occurs when you are able to use any dog hole in the bench. You can adapt the bolt to work in a ¾-inch dog hole, as shown in Illus. 5-15. Place a large fender washer between the two nuts. A fender washer has a much larger outside diameter than a standard washer.

To install the hold-down, insert the bolt from below the bench and hook the hold-down over the bolt head. If the hold-down has to be used in the same position for an extended time, put a cotter pin through the hole to lock the bolt in the slot.

Record makes a heavy-duty hold-down that is useful when you have to apply a lot of pressure or work with large parts (Illus. 5-16). To withstand the extra pressure this hold-down can exert, you must install a special socket in the bench top.

Illus. 5-16. This Record hold-down is useful for heavy-duty clamping.

BENCH STOPS

Sometimes you don't have to firmly clamp a board, but do have to prevent it from sliding out of position. Bench stops are designed for this situation. A bench stop (Illus. 5-17) is simply a thin strip of wood with two dowels attached that fit into dog holes in the bench top.

Bench stops are simple to make. Use hardwood to make them. They can be any thickness, but for most work ½-inch-thick stops are good because they won't interfere with the tools when you are using ¾-inch-thick lumber. Make the stop about 2 inches wide and long enough to reach two dog holes. You can make several sizes if you like, for example, a short one that fits across two adjacent holes, a longer one that fits across the full width of the bench, and an even longer one to place lengthwise on the bench.

To attach the dowels, drill two ¾-inch-diameter holes that line up with the dog holes. Stop drilling before you drill completely through. Cut two pieces from a ¾-inch-diameter hardwood dowel. Wipe some glue inside the hole in the stop and insert the dowels (Illus. 5-18). Then drill a pilot hole for a wood screw through the top of the board and into the dowel. Countersink the hole so that the screw head will be below the surface. Now, drive in a screw to secure the dowels in the holes.

Bench stops have many uses. I use them mostly for belt-sanding. I put a stop across the end of the bench and butt the end of the board I'm sanding against the stop. This prevents the board from slipping. A stop is much faster to use than clamps or the end vise, so when you have a stack of parts to sand you can save a great deal of effort by using a bench stop.

You can also use a bench stop when face-planing. Usually all you will need is a stop to prevent the board from sliding off the end of the bench. When you are hogging off a lot of wood with diagonal strokes (that is, making deep, rough cuts in wood with a plane), put another stop along the side of the board to prevent it from sliding sideways.

Sometimes I use a bench stop instead of a bench

Illus. 5-17. A bench stop is a thin strip of wood with two dowels attached that fit into dog holes in the bench top.

dog when clamping large parts with the tail vise. If you have a full-width tail vise, you can also put a stop in the dog holes of the vise. This lets you position the work anywhere on the bench instead of just aligning it with the dog holes.

A bench stop can also prevent parts from sliding on the bench when you are assembling a project. Once you have a few bench stops, you will find more and more uses for them.

WEDGES

Long before vises were invented, woodworkers used wedges to secure the work on a bench. Even today wedges can be a valuable addition to your bench. Wedges can speed up a job when you are working on many similar parts, because once they are set up, a tap with a mallet is all that is needed to secure or remove the work.

You can use the wedge bench stop shown in Illus. 5-19 to hold large parts flat on the bench top. To make a wedge bench stop, cut a board 4 inches wide. The length of the board depends on the width of your bench and the spacing between dog holes. Next, make a taper jig for your table saw (Illus. 5-20). Set the blade tilt to five degrees, and, using the taper jig, rip the board. This will produce a wedge and a fence with matching tapers. The five-degree saw angle makes a compound taper that prevents the wedge from lifting away from the bench top as you work. Install the dowels in the fence as described in the previous section.

To use the wedge bench stop to secure your work

Illus. 5-19. This wedge bench stop is useful for holding large parts flat on the bench top.

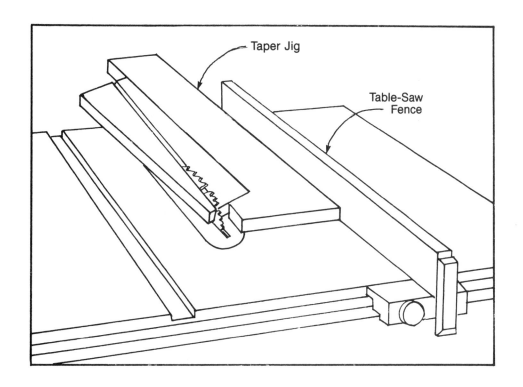

Taper Jig

Table-Saw Fence

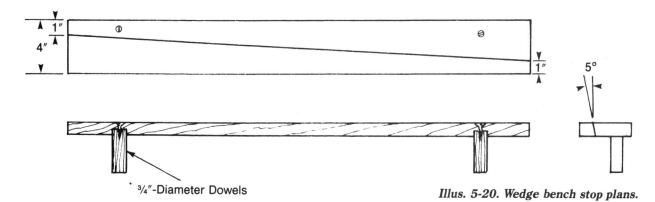

4″

1″

1″

5°

¾″-Diameter Dowels

Illus. 5-20. Wedge bench stop plans.

Illus. 5-21. Tap the wide end of the wedge with a mallet to clamp the work.

to the bench, place a standard bench stop on one edge of the work and the wedge bench stop on the other. The wedge bench stop fits in the dog holes like a standard bench stop. Tap the wide end of the wedge with a mallet to tighten it (Illus. 5-21). Tap the narrow end of the wedge to loosen it.

The wedge stop shown in Illus. 5-23 will hold a board edge-up on the bench. It is handy when you are edge jointing small boards. Make the wedge stop from hardwood. There are two variations shown in Illus. 5-22. One is permanently attached to the bench top with screws, and the other can be temporarily placed in the front vise.

BENCH HOOKS

A bench is meant to be used and a few nicks and gouges are inevitable, but it should not be abused. A bench hook (Illus. 5-24) is one way to protect your bench top when you are sawing or drilling.

Like a bench stop, when a bench hook is used, pressure from the tool instead of from a vise holds the work stationary. A bench hook hooks over the front edge of the bench and has a stop a few inches back from the edge. Bench hooks are useful for crosscutting small boards and trimming joints. A bench hook provides a firm stop to saw against, and it also protects the bench

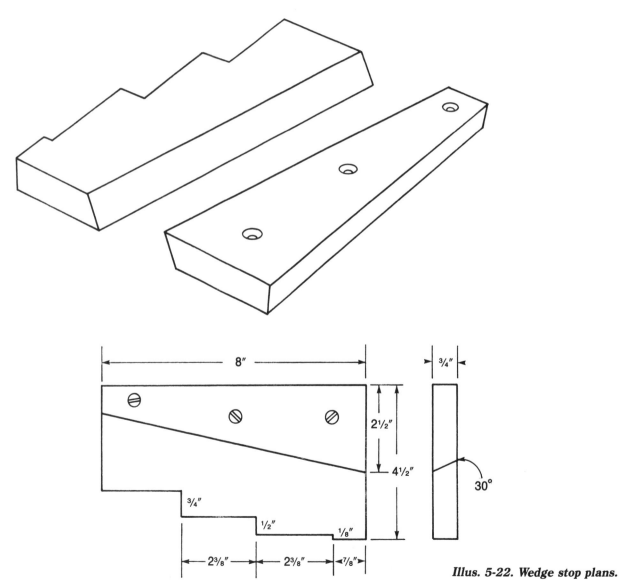

Illus. 5-22. Wedge stop plans.

Illus. 5-23. This wedge stop will hold a board edge-up on the bench. Here it is shown mounted to a board that can be clamped in the front vise. This wedge stop can also be attached permanently to the bench top with screws.

Illus. 5-24. A bench hook gives you a firm stop to saw against and protects the bench top from tool marks.

top from tool marks. When you are drilling holes, place the parts on a bench stop. You can use a hold-down to secure a board to the bench hook if necessary. When the drill breaks through the work, it will drill into the bench hook instead of the bench top.

A bench hook doesn't have to be fancy. It is simply a piece of wood, plywood or particleboard, with a strip of wood attached to the bottom face along the front and the top face along the back.

BOARD JACKS

When you are working on a long board clamped edge-up in the front vise, you must support the free end of the board. There are several types of board jacks that will do this. Illus. 5-25 shows a simple board jack. To use it, you must have ¾-inch dog holes drilled in the front edge of the bench top. It is simply a steel corner brace screwed to a length of ¾-inch dowel. A bullet

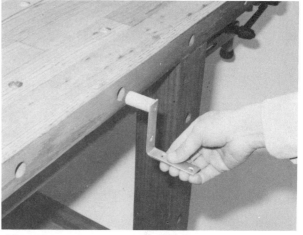

Illus. 5-25. You can make a simple board jack by attaching a length of ¾-inch dowel to a steel corner brace. The dowel fits in holes drilled in the front edge of the bench top. The corner brace supports the end of a long board clamped in the front vise.

catch in the side of the dowel will help to hold it in position. Choose a size of angle brace that allows you to place its bottom leg even with the top of the guide rods of the front vise.

For narrow boards, simply insert a length of dowel into one of the dog holes on the front of the bench. To handle very wide boards and things like cabinet doors, drill a row of ¾-inch holes down the front legs, and then insert a dowel into the holes for the board to rest on. This system works best with wide workbench legs that are flush with the front of the bench (Illus. 5-26). I used this design for the joiner's bench in Chapter Three. You can use a holdfast instead of a dowel in these holes. The holdfast will prevent the free end of a thin board from whipping as you work (Illus. 5-27). The Veritas hold-down will also fit in the board-jack holes (Illus. 5-28).

You can make a simple board jack to use with the classic cabinetmaker's bench described in Chapter Four. Cut a board about 2½ inches wide and 1½ inches thick. Make the board long enough to reach from the floor to the top of the bench. Now, drill a series of ¾-inch-diameter holes in the board. To use the board jack, rest one end on the floor and clamp the other in the tail vise. Insert a dowel and a holdfast or hold-down in one of the holes to support the board (Illus. 5-29).

Illus. 5-26. A dowel inserted in a hole drilled in the front leg of a workbench will support the end of a long board clamped in the front vise.

Illus. 5-27. A holdfast can also be used to support the end of a long board.

Illus. 5-28. Since the board-jack holes are ¾ inch in diameter, any bench accessory designed for ¾-inch holes will fit in the board-jack holes. This Veritas hold-down clamps the free end of a board securely against the bench.

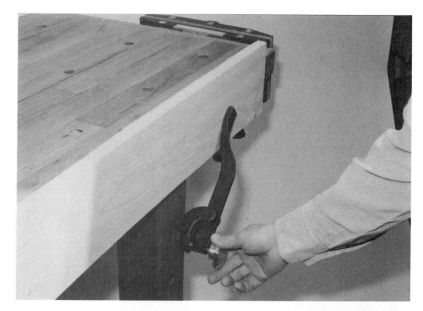

Illus. 5-29. You can make a board jack to use with a traditional tail vise by drilling a series of ¾-inch-diameter holes in a board and clamping it in the tail vise.

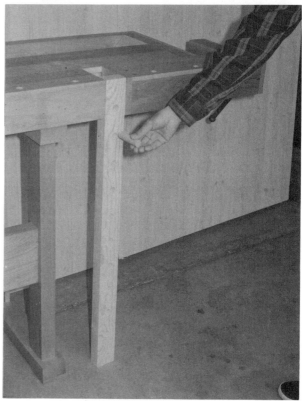

CHAPTER SIX

Using a Workbench

A WORKBENCH IS A very versatile tool. It will allow you to work more quickly and easily. There are many ways to use a workbench. In this chapter I describe how to use a workbench for planing, joinery, power-tool operations, and assembly.

PLANING

If you enjoy traditional woodworking with hand tools, you will find that a good workbench makes all planing operations faster and easier. In this section, I describe three planing operations: edge-jointing, face-planing, and sticking (using a moulding plane).

Edge-Jointing with a Hand Plane

To joint the edge of a board, you have to support the board edge up. The best way to do this is to use the front vise. To joint the edge of short boards, simply clamp them in the front vise (Illus. 6-1). A shoulder vise works well for this, because you can butt the end of the board against the support block (Illus. 6-2).

You can joint narrow boards clamped in a wedge stop (Illus. 6-3). A wedge stop is especially handy when you have many boards to plane, because you can release one board and secure the next one quickly.

For long boards, use the front vise and a board jack. The board jack on the joiner's bench gives you many options. You can put dowels in the holes along the front edge to plane narrow boards (Illus. 6-4). For wider boards, you can use the holes in the front legs of the workbench. The holes in the legs are also useful when you have to joint the edge of a door. Use two dowels to support the work, one in each leg. Rest the board on the dowels, and then slide the end of the board into the vise until it hits the guide rod. Clamp the vise tight on the end of the board. If the board moves excessively at its free end, use a hold-down instead of a dowel in the leg dog holes (Illus. 6-5).

Illus. 6-1. To joint short boards, simply clamp them edge-up in the front vise.

Illus. 6-2. A shoulder vise works well for edge-jointing.

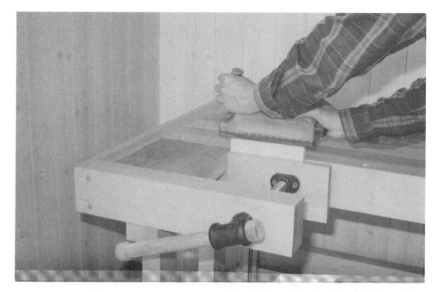

Illus. 6-3. A wedge stop is handy when you have to plane narrow boards.

Illus. 6-4. To joint the edge of longer boards, clamp the board in the front vise and support the board's free end with a board jack.

Illus. 6-5. If the board moves excessively at the free end, you can use a hold-down to clamp it in place.

The shoulder vise of the classic cabinetmaker's bench gives wide boards more support than other types of front vise, because there are no guide rods to get in the way. Clamp the board jack in the tail vise for planing long boards in the shoulder vise (Illus. 6-6).

Illus. 6-6. A shoulder vise gives wide boards more support than other types of front vise, because there are no guide rods to get in the way.

Planing a Board on its Face

You can use bench stops or the tail vise and bench dogs when planing the face of a board. To use bench stops,

place one at the far end of the bench and one or two along the edge of the board. Place the board against the stops. The forward pressure from the plane will force the work against the stops and secure the board (Illus. 6-7). The stops along the edge are necessary when you take diagonal cuts across the face in the early stages of planing (Illus. 6-8). If you are simply smoothing a board that has already been planed to the correct thickness, you only need the stop at the end. If you are planing correctly, all of the force will be in a straight line against the bench stop.

The tail vise will hold a board more securely than a bench stop. This can be an advantage when you are planing boards with difficult grain. To secure a board for planing with the tail vise, lay the board on the bench with its end resting on top of the tail vise. Select dog holes in the vise and the bench top that will let you clamp the board with the smallest vise opening. Insert the dogs and adjust their height so that the top of the dog is about ¼ inch below the top face of the board. Now, tighten the vise to clamp the board tight (Illus. 6-9).

A bench with two or more rows of dog holes and a wide tail vise can be useful when you have to plane a wide board. The second set of dogs holds a wide board more securely (Illus. 6-10).

Sticking

The tail vise and a bench dog can hold a board stationary without interfering when you are sticking, that is, using a moulding plane. If you are using a plane with a fence, the front edge of the board should overhang the edge of the bench top slightly so that the fence

Illus. 6-7. You can prevent a board from moving as you plane its face by butting its end against a bench stop.

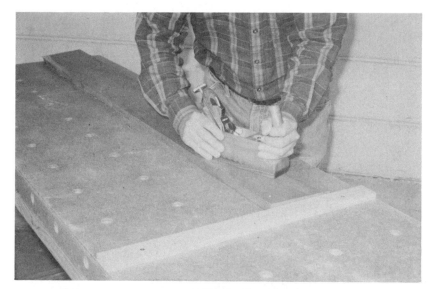

Illus. 6-8. When you are hogging off a lot of wood, add a second bench stop along the edge of the board to prevent it from moving as you take diagonal cuts across its face.

Illus. 6-9. You can clamp a board securely for face-planing with a tail vise and bench dogs.

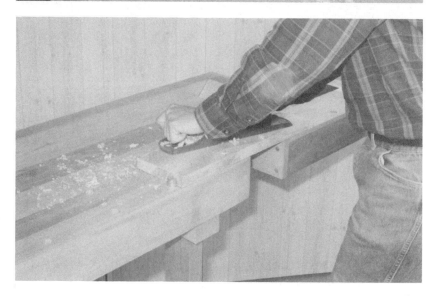

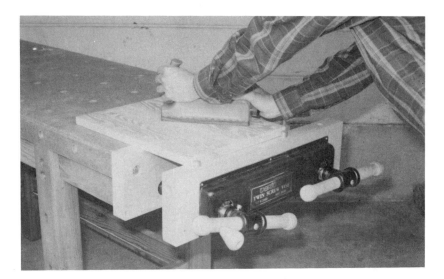

won't hit the bench top. The classic cabinetmaker's bench is designed specifically for cabinetmaking with hand tools, so the dog holes are close to the front edge. With this bench you can clamp a board as narrow as 1¾ inches with its edge slightly overhanging the bench top (Illus. 6-11).

Illus. 6-11. Dog holes close to the front edge of the classic cabinetmaker's workbench make it possible to clamp narrow boards for sticking.

The joiner's bench is designed for larger work, so the dog holes are spaced farther from the front edge. Using the standard dog holes, you can stick a board as narrow as 4 inches (Illus. 6-12). If you have to stick a narrower board on the joiner's bench, use bench stops. You will need two bench stops that reach to the edge of the bench. Insert one bench stop in the tail-vise dog holes and the other in a set of dog holes in the bench top. This arrangement allows you to clamp very narrow boards with their front edges slightly overhanging the benchtop (Illus. 6-13).

JOINERY

A good workbench helps you make accurate joints, because it holds the work firmly and allows you to use both hands on the tools for good control. In this section, I describe how to secure work while mortising, making tenons, and dovetailing.

Mortising

You can use a hold-down to secure a small part such as a table or chair leg to the bench top while you cut mortises. Place the part near one of the workbench legs. This is the steadiest part of the bench top, so there won't be any movement as you chop the mortises. Place a piece of scrap wood between the hold-down and the work, to prevent the hold-down from marring the work (Illus. 6-14). Usually a hold-down will be all you need to prevent the work from slipping; however, you can place a bench stop alongside the work to steady it, and then clamp it with a hold-down.

You can also use the tail vise to secure work for mortising (Illus. 6-15). Always make sure that the end

Illus. 6-12. The joiner's bench dogs are not as close to the front edge, so a board must be at least 4 inches wide for you to be able to clamp it with its edge overhanging the bench top.

Illus. 6-13. Narrower boards can be clamped with their edges overhanging the bench top if you use bench stops instead of dogs.

Illus. 6-14. You can use a hold-down to secure the work for mortising.

Illus. 6-15. When you clamp a board in a tail vise for mortising, make sure that you chop the mortise on the bench top, not the tail vise.

to be mortised is on the bench top, not the tail vise. The tail vise is not designed to withstand that much downward pressure. The classic cabinetmaker's bench is designed so that there is a leg directly below the inside vise jaw. This provides extra strength for mortising.

A wedge stop is especially good for holding boards edge up for mortising (Illus. 6-16). The wedging action will prevent the board from tipping over as you chop the mortise.

Illus. 6-16. A wedge stop can also be used to hold a board while you chop a mortise.

Cutting Tenons

The shoulder vise is ideal for cutting tenons, because there are no guide rods to get in the way. Align the board with the vise screw (Illus. 6-17).

A face vise with guide rods can also be used when tenons are being cut. A large vise with several inches of clearance between the guide rod and the edge of the jaw is best (Illus. 6-18).

A twin-screw vise is very good for cutting tenons on wide boards. Place the work between the two screws (Illus. 6-19).

Illus. 6-17. A shoulder vise is ideal for cutting tenons, because there are no guide rods to get in the way. The jaw also adjusts to fit odd-shaped parts. Notice that the jaw is slightly angled to conform to the taper of this table apron.

Illus. 6-18. Cutting a tenon in an iron face vise.

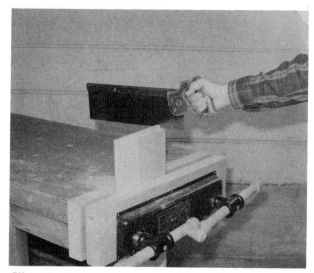

Illus. 6-19. A twin-screw vise is good for holding wide boards when you cut tenons.

Cutting Dovetails

When you saw the pins and tails of a dovetail joint, the board must be held securely end up. The shoulder vise is ideal for this operation, because it doesn't have any guide rods to get in the way (Illus. 6-20).

Illus. 6-20. Cutting dovetails on a board held in a shoulder vise.

The twin-screw end vise can also hold wide boards end up as you saw the pins and tails of a dovetail joint (Illus. 6-21). You can place a board as wide as 13 inches between the vise screws of a twin-screw vise mounted in the tail position on the joiner's bench. If you mount the vise in the front position, you can space the vise screws even farther apart.

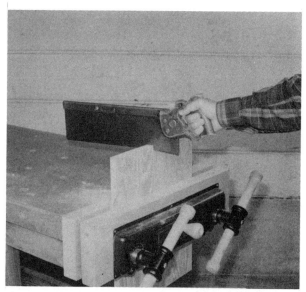

Illus. 6-21. Cutting dovetails on a board held in a twin-screw vise.

You can clamp boards in an iron vise or a wooden vise that has guide rods, but one edge will be unsupported. This is not a problem with boards up to about 6 inches wide (Illus. 6-22). For wider boards, you can use a bar clamp across the bench top to secure the other edge of the board (Illus. 6-23).

A hold-down can be used to secure the work on the bench top as you chop out the waste with a chisel (Illus. 6-24). Chop halfway through the boards, and then turn them over to complete the cut.

Illus. 6-22. Cutting dovetails in an iron vise.

Illus. 6-23. You can cut dovetails on a wide board in an iron vise by securing the free end of the board with a bar clamp.

Illus. 6-24. You can use a hold-down to secure the work as you chop out the waste on a dovetail joint.

POWER-TOOL OPERATIONS

A good workbench can make it simpler and safer to perform many power-tool operations. The bench will hold the work securely, so you can use both hands to guide the tool. Ordinary clamps often get in the way when you are using power tools. The dog holes in a workbench provide several ways to clamp the work without interfering with the tool. Below I describe ways to secure the work when belt-sanding and routing.

Belt-Sanding

You can use the tail vise and bench dogs to secure the work when belt-sanding (Illus. 6-25). The tail vise will hold the work securely, so it is particularly useful with long, narrow boards.

The large, flat surface of the joiner's bench makes it an ideal bench for belt-sanding wide boards. You can

Illus. 6-25. You can use a tail vise and bench dogs to hold a board securely when belt-sanding.

use the tail vise and dogs or bench stops. Bench stops work well when you are belt-sanding, because all of the force is pushing the board rearward. A stop at the end of the bench is all that is necessary to keep the board stationary (Illus. 6-26). A bench stop is very convenient when you have many boards to sand, because you don't have to loosen and tighten a vise each time you turn a board over or change to another board.

When sanding a long, narrow board, you may want to place another stop along its edge to prevent any side-to-side movement (Illus. 6-27).

Routing

It is important to hold the work securely when routing. C-clamps and hand screws often get in the way when you are routing, so you must reposition them as you work. The tail vise, dog holes, and bench accessories provide several options for holding the work while you are routing without interfering with the router base.

When routing the edge of a board, make sure that the edge overhangs the front of the bench slightly. You can do this using the tail vise and bench dogs. Position the dogs about ¼ inch below the top surface of the board. With this setup, there are no projecting clamps to interfere with the router as you make the cut (Illus. 6-28). The joiner's bench is especially good for holding large parts for routing. For wide boards, use both dog holes in the end vise and two dogs in the bench top (Illus. 6-29).

When making a cut on the face of a board, you can hold the board with the tail vise and dogs or hold-downs. Hold-downs can also be used to clamp a guide board to the work to guide the router for operations such as cutting dadoes (Illus. 6-30).

Illus. 6-26. Using a bench stop is a convenient way to prevent a board from moving rearward while you belt-sand.

Illus. 6-27. When belt-sanding a long, narrow board, you may have to place another stop along its edge.

Illus. 6-28. When you clamp a board with a tail vise and bench dogs, there is nothing projecting above its surface to interfere with a router.

Illus. 6-29. A twin-screw vise is good for holding wide boards when you are routing.

Illus. 6-30. You can use hold-downs to clamp a guide board when making dadoes with a router.

Odd-shaped parts can be secured with Veritas Wonder Dogs (Illus. 6-31). Hold-downs can also be used to secure odd-shaped parts when they won't interfere with the router.

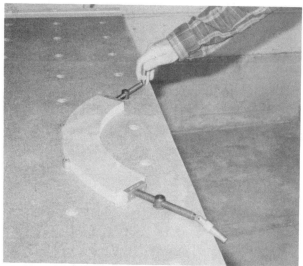

Illus. 6-31. Veritas Wonder Dogs can be used to clamp parts that would be difficult to secure by other means. Notice that there are no obstructions on the inside curve of this board, where routing will take place.

TEMPORARILY MOUNTING TOOLS AND JIGS TO THE BENCH

You can quickly mount tools and jigs to the bench using hold-downs (Illus. 6-32). If the tool or jig doesn't have a good clamping surface, you can make a plywood or particleboard base for the tool. For added stability, attach ¾-inch dowels to the base that are aligned with the dog holes in the bench top (Illus. 6-33). The dowels fit into dog holes in the bench top. For some operations, all you have to do is insert the dowels in the dog holes. If you have to hold the tool or jig more securely, make the base large enough so that you can also clamp it with hold-downs.

ASSEMBLING PROJECTS

In a large shop, it is a good idea to have a separate assembly table. This allows you to keep the workbench clear for trimming and fitting parts as you assemble the project. In a small shop, the workbench must usually double as an assembly table.

The joiner's bench makes a good assembly table. If you are making a separate assembly table, follow the plans for the joiner's bench supplied in Chapter 3, but make it less than a normal bench height; 27 inches is a good height for an assembly table. You don't have to mount vises on an assembly table, but the dog holes will still be useful.

If you are careful as you drill the dog holes and make the bench stops, you can use two bench stops to square projects as you assemble them. Place the bench stops as shown in Illus. 6-34. Check them with a framing square to make sure that they are at a 90-degree angle to each other. Now, slide parts of a project into the bench stops to square them.

Wonder Dogs can be used instead of bar clamps to clamp small parts (Illus. 6-35). This is an excellent way to make sure that you are clamping a door flat. The bench top is a flat surface so if the frame is resting flat on the bench top as the glue dries, it should remain flat after you remove it from the clamps. Put some waxed paper or plastic wrap on the benchtop to prevent the glue from sticking to the bench top.

Illus. 6-32. Tools and jigs can be quickly mounted to the bench with hold-downs.

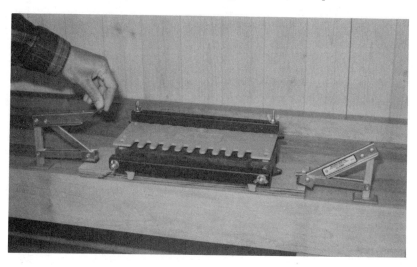

Illus. 6-33. For frequently used jigs, you can add a base with ¾-inch dowels that fit into the dog holes.

Illus. 6-34. If you lay out the dog holes carefully, you can use bench stops to help square your projects.

Illus. 6-35. Wonder Dogs can be used to clamp joints for gluing. Here I'm repairing a small panel door.

CHAPTER SEVEN

Workbench Storage Space

I N A SMALL shop storage space is usually limited, so you have to make use of every available storage area. In this case, you can use the workbench base for tool storage. In larger shops, you can build separate tool-storage cabinets and leave the workbench base open.

The Shakers usually designed their workbenches with storage cabinets in the base. The traditional European workbench has an open base. There are advantages to both types. Drawers and cabinets in the workbench base keep tools close to the work, so they are always handy when needed. However, boards clamped in the vise or overhanging parts of a project can block a drawer or cabinet door, making it impossible for you to get to the tool needed without moving the work. An open base provides plenty of room for your legs and feet while you are standing, and you can sit comfortably on a stool and work on small parts.

In this chapter, I describe how to build a workbench base that includes drawers and a cabinet with a door. This base can be used with a variety of bench tops. In Illus. 7-1, it is shown with a commercially made laminated maple top. You can also use the built-up particleboard-and-hardboard top described in Chapter Three. If you add the optional fifth leg, you can use the bench top described in Chapter Four.

Illus. 7-1. In this chapter, you will learn how to build this workbench base that includes drawers and a cabinet.

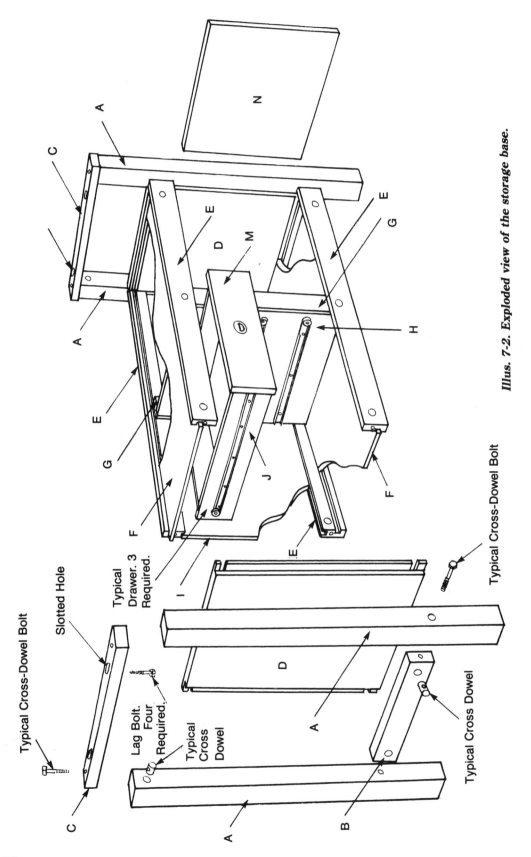

Illus. 7-2. Exploded view of the storage base.

Typical Cross-Dowel Bolt

Slotted Hole

Typical Cross-Dowel Bolt

Typical Drawer. 3 Required.

Lag Bolt. Four Required.

Typical Cross Dowel

Typical Cross Dowel

Typical Cross-Dowel Bolt

POST-AND-PANEL CONSTRUCTION

To add cabinets to the base of a workbench, you have to enclose the base. The strongest enclosed base is built with post-and-panel construction. This method maintains the strong base design of an open-base bench and fills in the areas between the frame members with thin panels. The panels can be made of plywood, hardboard, or solid wood.

You can adapt most base designs to post-and-panel construction. The only change in design is a groove around the inside of each opening. The plans presented

Storage Base Bill of Materials

		Description	Size (Inches)	Material	Number Required
A		Legs	2¼ × 2¼ × 32	Maple (Glue Three ¾" Thick Boards Together)	4
B		Stretcher	1¼ × 2¾ × 14½	Maple	2
C		Top Rail	1¼ × 2¼ × 19	Maple	2
D		Side Panel	¾ × 16 × 20½	¾" Plywood	2
E		Rails	¾ × 2¾ × 36½	Maple	4
F		Top and Bottom Panels	¾ × 16¾ × 36½	¾" Plywood	2
G		Mullion	¾ × 2¾ × 18	Maple	2
H		Center Panel	¾ × 15¼ × 18	¾" Plywood	1
I		Back Panel	¼ × 15¾ × 37¼	¼" Plywood	1
J		Drawer Sides	½ × 5 × 15	Maple	6
K		Drawer Back	½ × 4¼ × 14¾	Maple	3
L		Drawer Bottom	¼ × 14¾ × 15	¼" Plywood	3
M		Drawer Front	¾ × 5 × 17¼	Maple	3
N		Door	¾ × 15 × 17¼	¾" Plywood	1

Storage Base Hardware List

Number Required	Description
16	¼" × 4" Bolt-and-Cross-Dowel Fasteners
16	¼" Washers
4	#12 × 1" Panhead Wood Screws
4	#12 Washers
3 sets	15" Steel Drawer Guides
4	Flush Pulls
12	Drawer Fasteners (optional)

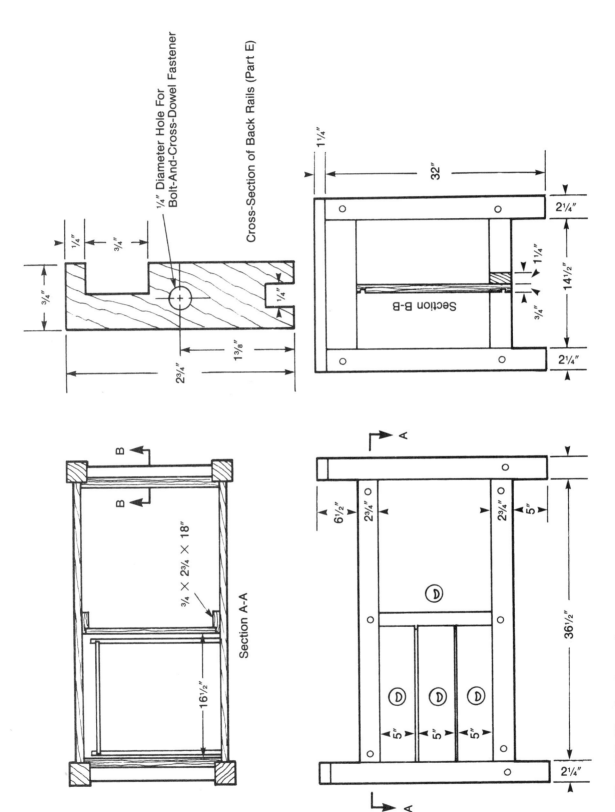

¼" Diameter Hole For
Bolt-And-Cross-Dowel Fastener

Cross-Section of Back Rails (Part E)

Section B-B

Section A-A

¾ × 2¾ × 18"

Illus. 7-3. Plan for storage base.

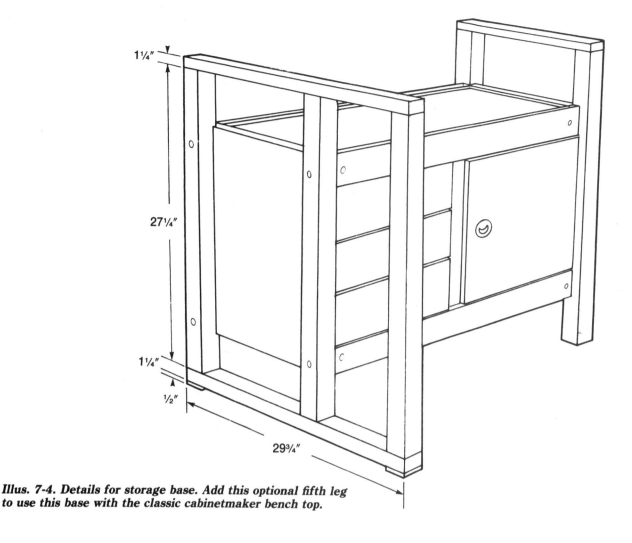

1¼″

27¼″

1¼″

½″

29¾″

Illus. 7-4. Details for storage base. Add this optional fifth leg to use this base with the classic cabinetmaker bench top.

in this chapter (Illus. 7-2–7-4) use a modified post-and-panel construction.

The back, top, and bottom panels for the workbench base featured in this chapter fit in grooves. The side panels are made from ¾-inch-thick plywood. Instead of fitting in grooves in the legs, they are attached to the inside of the legs with screws. The plywood sides make the base more rigid and provide a solid mounting surface for the drawer guides. This system works well with plywood, but don't use it with solid lumber.

The easiest way to make the grooves is with a router. A plunge router is helpful, because you have to make a plunge cut at the beginning of some of the grooves (Illus. 7-5). Install a ¼-inch straight bit in the router and adjust it to make a ⅜-inch-deep cut. Use a fence to guide the router. Adjust the fence to center the groove. You can make the groove along the full

Illus. 7-5. Use a router to cut the grooves in the legs of the workbench base.

length of the stretchers. Make blind grooves in the legs. Mark the location of the stretchers on the legs, so you will know where to start and stop the cuts.

Assemble the ends of the base first. The end panels are made of ¾-inch-thick plywood, so that you can screw the drawer guides to the panels. The panels are rabbeted on both edges.

After the frame is assembled, attach the panels to the inside of the frame with screws. Next, add the stretchers and the rear panel. The rear panel is made of ¼-inch-thick plywood. It fits into a groove in the stretchers and the legs (Illus. 7-6).

Illus. 7-6. The rear panel fits into a groove in the stretchers and legs.

Bolt-and-cross-dowel-fasteners are used to attach the stretchers to the legs (Illus. 7-7). You can buy commercial bolt-and-cross-dowel-fasteners from mail-order woodworking-supply companies or make your own as described in Chapter Four. Drill the holes for them using a drill guide or a drill press.

Once you have the base assembled, attach the top. The top is held in place by four ⅜-inch lag bolts. Notice that Illus. 7-2 shows slotted bolt holes. The elongated hole allows the top to shrink or swell as necessary. Be sure to use an additional ½-inch washer on each bolt

Illus. 7-7. Bolt-and-cross-dowel fasteners are used to attach the stretchers to the legs.

under the ⅜-inch washer. The larger washer prevents the bolt head from digging into the wood.

DOOR

There are three types of door you can use on a workbench: flush, lipped, and overlay (Illus. 7-8). Flush doors are good for workbenches, because the door front doesn't extend past the frame members. With some bench designs, a door that overlaps the frame may interfere with boards in the vise. Flush doors are easy to make, but you must fit them carefully. If you don't, the gap around the door will show.

A lipped door will hide the gap, so it is easier to fit. The lip can be cut with a router or a dado blade on a table saw. The standard size for the lip is ⅜ inch. You must use special hinges designed for a lipped door.

An overlay door is easier to make than a lipped door. It is also easy to install, because it doesn't fit inside the opening. This is a good type to use if the bench design allows it. However, the doors will extend ¾ inch past the frame, so they may interfere with boards in the vise if the frame is close to the front edge of the bench top.

The base shown in Illus. 7-2 has a flush door made from ¾-inch-thick plywood. Simply cut the door to size and sand the edges. Screws don't hold well in a plywood edge, so use hinges that allow you to drive the screws into the rear face of the door (Illus. 7-9). After the door is in place, install a magnetic door catch.

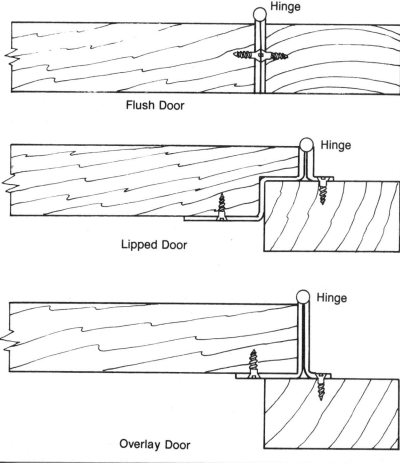

Illus. 7-8. Three types of door can be used on a workbench: flush, lipped, and overlay.

Flush Door

Hinge

Lipped Door

Hinge

Overlay Door

Hinge

Illus. 7-9. Screws don't hold well in a plywood edge, so use these special hinges with a bent leaf that allow you to drive the screws into the rear face of the door.

DRAWERS

Drawers are handy for storing small hand tools. Drawers for a workbench have to be especially sturdy because tools are heavy. A drawer design that works fine in a kitchen cabinet will soon break under the load of heavy tools. Use ¼-inch-thick plywood or hardboard

for the drawer bottom and at least ½-inch-thick wood for the drawer sides.

Illus. 7-10 shows two ways to build the drawers. Assembling the drawers with commercial fasteners is a simple way well suited for beginning woodworkers (Illus. 7-11). You can buy these fasteners at a well-stocked hardware store or from a mail-order wood-

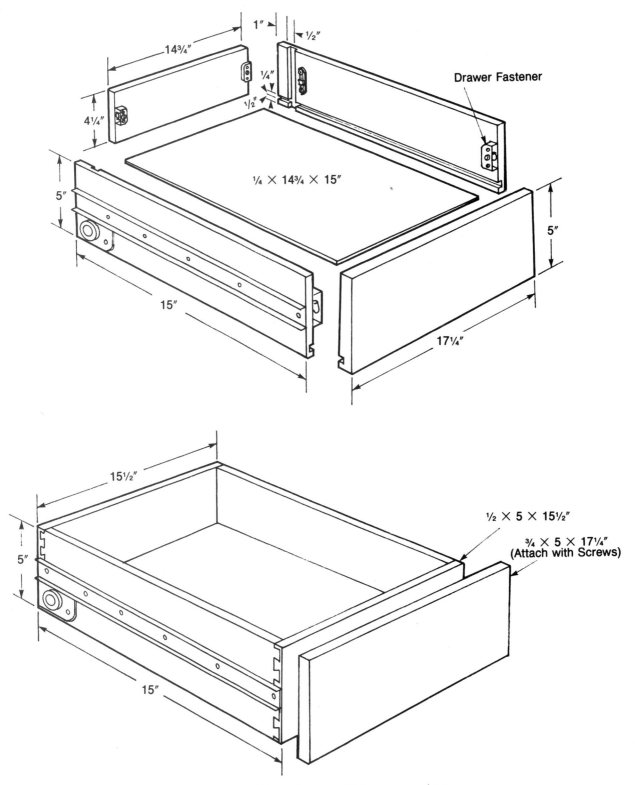

Illus. 7-10. This plan shows two ways to build the drawers. Using commercial fasteners is a simple way for beginners to build the drawers. Dovetailed drawers are stronger.

Illus. 7-11. These commercial fasteners simplify drawer construction.

working-supply store. Once the two halves of the fastener have been attached to the parts, lock the joint together by turning the center cam with a screwdriver.

The dovetail joint is the strongest drawer joint. If you are a more experienced woodworker, perhaps you should use dovetail joints. You can buy an inexpensive jig that will guide a router to cut the dovetails (Illus. 7-12).

The drawer guides have to be strong to handle the weight of the tools in the drawer. The plans specify steel drawer guides. Steel drawer guides are strong and easy to install. They also have rollers so the drawer will always operate smoothly and a built-in stop so you

won't accidently pull the drawer all the way out and dump your tools on the floor.

Choose a heavy-duty steel drawer guide. Buy the drawer guides before you begin making the drawers, because the exact size of the drawers depends on the clearance that the drawer guides require. Most drawer guides have a standard ½-inch clearance on each side. The dimensions shown in Illus. 7-10 and in the Bill of Materials allow for a ½-inch drawer guide clearance on each side of the drawer.

Use hardwood lumber for the drawers. The fronts are ¾ inch thick and the sides and backs are ½ inch thick. The bottom is made from ¼-inch-thick plywood or hardboard.

Begin by cutting all of the parts to size, and then cut the groove for the bottom with a router and a ¼-inch straight bit. If you have already assembled the base and attached the vises, clamp the drawer parts between a bench dog and the tail vise. Make sure that the dogs are slightly below the face of the work, so they won't interfere with the router base. Let the bottom edge overhang the edge of the bench slightly, so that there will be clearance for the router fence. Now, place the router on the work and make the cut (Illus. 7-13).

If you will be using commercial drawer fasteners, make the dado at the rear of the drawer next. Clamp the drawer side to the bench top with a hold-down and use a ½-inch bit in the router to make the cut. Guide the router with a router fence.

If you are using a dovetail construction option, use the router and a dovetail jig to cut the dovetail joints. Mount the dovetail jig on a board and clamp the board

Illus. 7-12. This inexpensive jig will guide a router to cut the dovetails.

Illus. 7-13. Use a router to cut a groove for the drawer bottom.

to the bench with hold-downs. Position the jig so that the front clamp overhangs the edge of the bench.

Now, clamp the drawer parts in the dovetail jig and cut the joints following the directions that come with the jig (Illus. 7-14).

If you like, round over the edges of the drawer fronts with a rounding-over bit in a router. Now, assemble the drawers. You can use bench stops to help square the drawers. Place one stop in the back row of dog holes and one across the end. Check the stops with a framing square. If you drilled the holes accurately, they should be square with each other.

Apply glue to the dovetail joints and drive them together with a mallet. Now, put the front of the drawer against the end bench stop and one side of the drawer against the rear bench stop. Push the drawer against the stops until it is resting flat against both of them. Now, the drawer is square.

Slide the drawer bottom in from the rear. Push the drawer against the bench stops and push the drawer bottom all the way into the groove. Drill pilot holes for the two screws that secure the bottom of the drawer to its back. Now, drive the screws as you hold the drawer against the bench stops. With its bottom secured, the drawer will stay square. If the dovetail joints are tight enough, you may not need any clamps. If they won't stay tight by themselves, use bar clamps to hold them as the glue dries.

PULLS

Door and drawer pulls can be a problem on a workbench. They can interfere with the work (Illus. 7-15) and you can get your clothes caught on them while working. If you use a pull that extends out from the surface, choose a small pull with rounded edges. The best type of pull is a flush pull (Illus. 7-16). You can

Illus. 7-14. This dovetail jig cuts both parts of the joint at once.

buy several commercial types of flush pull. You can eliminate the need for a pull by cutting a handhold (Illus. 7-17). The only disadvantage of a handhold is that it lets more dust into the drawers than a closed front.

Illus. 7-15. If you use drawer pulls that stick out too far, they can interfere with boards clamped in the vise.

Illus. 7-16. Flush pulls won't get in the way.

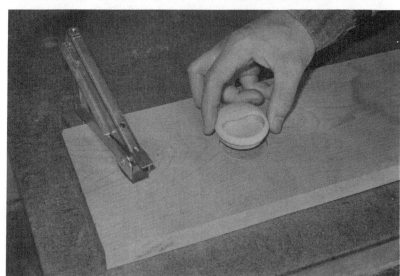

Illus. 7-17. You can eliminate the need for a pull by cutting a handhold with a sabre saw.

CHAPTER EIGHT

Outfitting Your Bench for Specialty Work

ALL OF THE benches described in this book are versatile enough to be used for any woodworking project, but if you specialize in one area of woodworking you may find that some combinations of bench and accessories are better suited for the work you do than others. In this chapter, I describe what types of bench and accessories will prove best for you.

GENERAL WOODWORKING

If you want a bench that works well for a variety of work, I recommend a large joiner's bench. Equip the bench with a good-quality iron front vise (Illus. 8-1). You can also add a second iron vise in the tail vise position (Illus. 8-2). Drill the complete grid of dog holes and buy several hold-downs. Black & Decker hold-downs are inexpensive and work well for general woodworking (Illus. 8-3).

FURNITURE MAKING

The classic cabinetmaker's bench is specifically designed for furniture making (Illus. 8-4), so it is the bench many furniture makers prefer. Another good bench for furniture making is a large joiner's bench with an iron front vise and a twin-screw tail vise (Illus. 8-5). The twin-screw tail vise is very helpful when you are working on large carcass parts.

Illus. 8-1. A large joiner's bench with a good-quality iron front vise is a good choice for general woodworking.

Illus. 8-2. A second iron vise in the tail vise position works well for general woodworking.

Illus. 8-3. Black & Decker hold-downs work well for general woodworking.

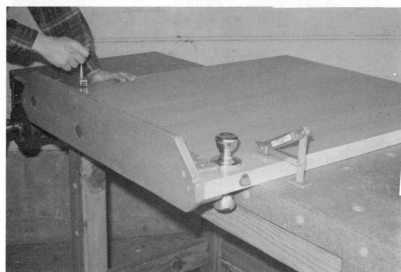

Illus. 8-4. The classic cabinet-maker's bench is the type of bench many furniture makers prefer.

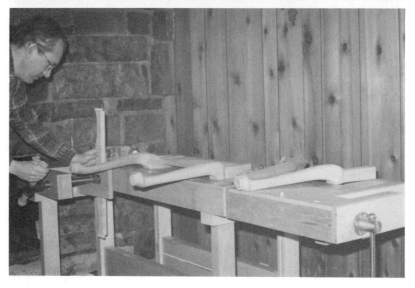

Illus. 8-5. A large joiner's bench with a twin-screw tail vise is very useful when you are working on large carcass parts.

FURNITURE-MAKING ACCESSORIES

Hold-downs are very useful for furniture making. You probably should have at least two. The Veritas hold-down is good for furniture making because of its capacity for clamping large projects and its strong grip (Illus. 8-6). The Record hold-down also has a large clamping capacity and strong grip (Illus. 8-7).

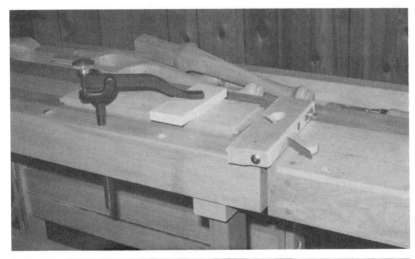

Illus. 8-6. The Veritas hold-down is a good choice for furniture makers. It fits into a ¾-inch dog hole.

Illus. 8-7. The Record hold-down is another good choice for furniture making. It fits into an iron socket in the bench top.

Poppets are useful if you will be making cabriole legs. They allow you to work on all areas of the leg conveniently (Illus. 8-8).

Any hold-down will secure a board to the bench as you chop dovetails, but if you frequently make hand-cut dovetails, you might want to make a special dovetail hold-down (Illus. 8-9). The hold-down fits in the dog holes in the bench top (Illus. 8-10). Make the dovetail hold-down from a hard wood like maple. Follow the plans shown in Illus. 8-11.

Illus. 8-8. Poppets are useful when you have to work on turned parts like this chair leg.

Illus. 8-9. This dovetail hold-down fits into the dog holes in the bench top. You can buy the threaded knobs from a mail-order woodworking supply company or you can substitute large wing nuts.

Illus. 8-10. To install the dovetail hold-down, insert the threaded rod into the dog holes from the underside of the bench top. Large fender washers compensate for the oversize hole.

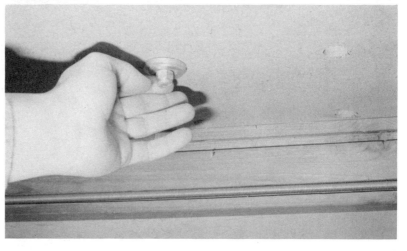

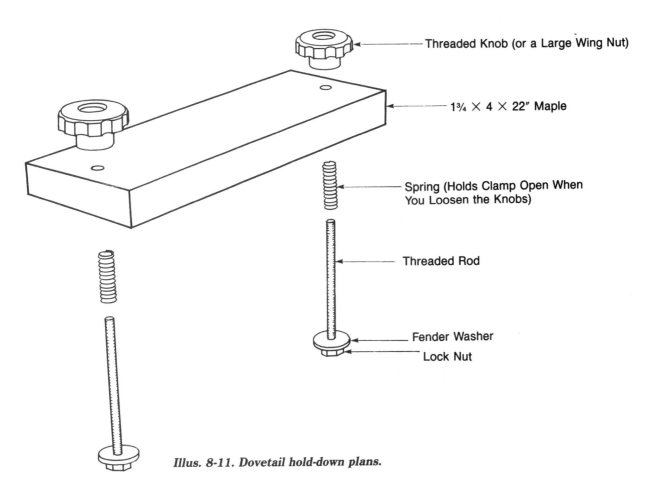

Threaded Knob (or a Large Wing Nut)

1¾ × 4 × 22″ Maple

Spring (Holds Clamp Open When You Loosen the Knobs)

Threaded Rod

Fender Washer

Lock Nut

Illus. 8-11. Dovetail hold-down plans.

CARVING

If you will be using your bench exclusively for wood carving, you probably won't need a large bench, but you will need a very sturdy one. A joiner's bench with a 24 × 58-inch top makes a good carver's bench (Illus.

Illus. 8-12. A joiner's bench with a 24 × 58-inch laminated top makes a good carver's bench.

Illus. 8-13. A bench top with a tool tray like that on this classic cabinetmaker's bench is handy for wood-carvers because they can keep sharpening stones and frequently used tools close to the work.

8-12). If you want to keep your carving tools handy, add a tool tray to the bench top (Illus. 8-13). You can also use the storage base described in Chapter Seven (Illus. 8-14).

Some wood-carvers prefer to sit down as they carve (Illus. 8-15). Illus. 8-16 is a plan for a sturdy bench stool.

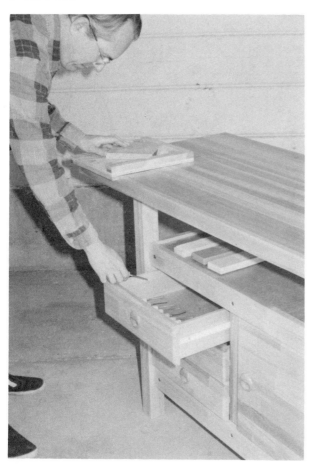

Illus. 8-14. The storage base described in Chapter 7 works well in a carver's bench because you can keep tools close at hand.

Illus. 8-15. Some wood-carvers prefer to sit down as they carve. This bench stool is comfortable and sturdy.

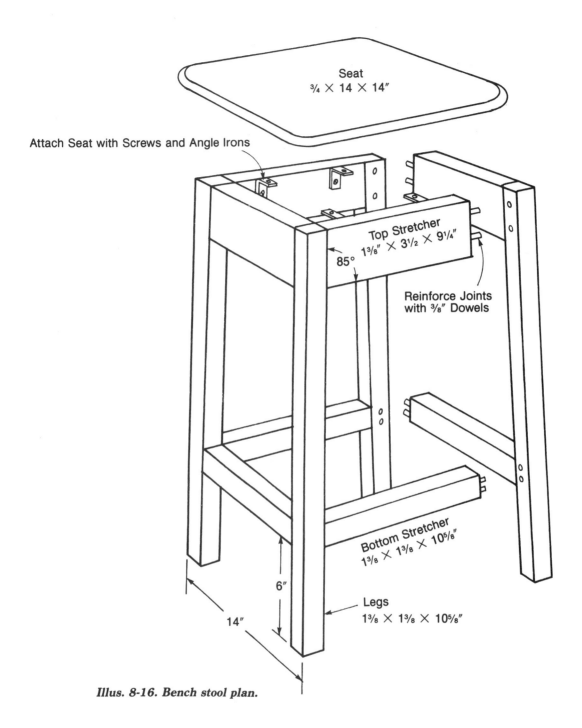

Seat
¾ × 14 × 14″

Attach Seat with Screws and Angle Irons

Top Stretcher
1⅜″ × 3½ × 9¼″

85°

Reinforce Joints
with ⅜″ Dowels

Bottom Stretcher
1⅜ × 1⅜ × 10⅝″

6″

Legs
1⅜ × 1⅜ × 10⅝″

14″

Illus. 8-16. Bench stool plan.

CARVING IN THE ROUND

For carving in the round, that is, carving all sides of a project, you have to be able to get at all sides of the work easily and you may have to reposition the work frequently. There are a few ways to accomplish this. A *Tucker vise* is good for wood carving, because you can reposition the work easily to any angle (Illus. 8-17). Many carvers prefer to mount the carving securely to the bench with a *carver's screw*. This is a long screw that has wood threads like a lag screw on one end and machine threads on the other (Illus. 8-18). You must drill a pilot hole in the bottom of the carving for the screw. Drive the lag end of the screw into the pilot

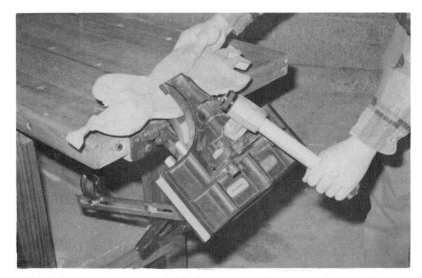

Illus. 8-17. For carving in the round, a Tucker vise is convenient to use. You can reposition the vise to hold the work at the most comfortable working angle.

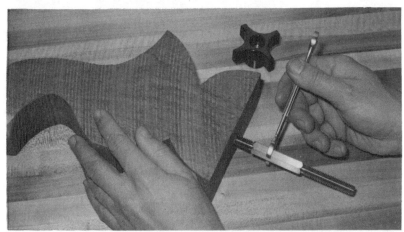

Illus. 8-18. This carver's screw has tapered wood threads on one end and machine threads on the other. Drill a pilot hole in the bottom of the carving before driving the screw.

hole until all of the lag threads are inside the wood; then place the machine screw end into a dog hole in the bench top. Place a large washer over the screw and then the nut. A carver's screw will usually come with a large wing nut or a knob; tighten the nut to secure the work (Illus. 8-19).

For small work that you must reposition frequently, you can use a *rope hold-down*. This is simply a length of rope looped through a dog hole and tied to a wooden foot pedal (Illus. 8-20 and 8-21). The work fits between the rope and the bench top. As long as you maintain pressure on the foot pedal, the work is held firmly, but when you want to reposition the work, just release the pedal and the work is free to move. Illus. 8-22 shows a single-loop rope hold-down. Illus. 8-23 shows a double-rope version used for larger work.

Illus. 8-19 (right). After the carver's screw has been driven into the carving, insert it in a dog hole and attach it to the bench top with the large wing nut or knob that comes with the carver's screw.

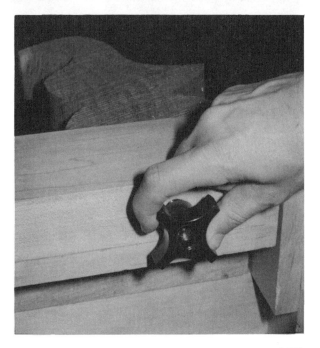

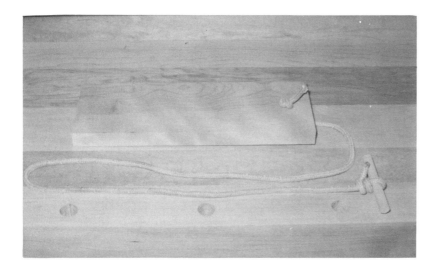

Illus. 8-20. A rope hold-down is simply a length of rope tied to a short dowel at one end and a board at the other. The board serves as a foot pedal.

Illus. 8-21. As long as you maintain pressure on the foot pedal, the rope hold-down will secure the work. To reposition the work, release the foot pedal.

Illus. 8-22. The single-loop rope hold-down has one loop through one dog hole. The dowel tied to the end keeps the rope from pulling through the dog hole.

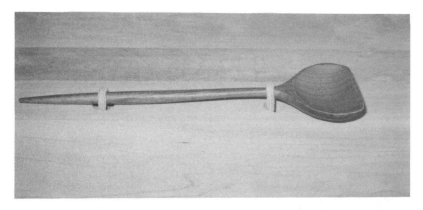

Illus. 8-23. A double-loop rope hold-down has a longer rope than a single-loop rope hold-down. Both ends of the rope are attached to the foot pedal. Fit its loop through two different dog holes.

PANEL CARVING

For panel carving, you can secure the work with a tail vise and dogs (Illus. 8-24). Many carvers prefer to use a bench hook, because it allows them to reposition the work quickly (Illus. 8-25). When using a bench hook, reposition the work each time you change the direction of your carving strokes so that the strokes always push the work against the stop. Illus. 8-26 shows a bench hook specifically designed for wood carving. It has several notches to help hold irregularly shaped work.

Illus. 8-24. When you are carving a large, flat panel, a tail vise and dogs will hold the panel securely.

Illus. 8-25. When carving small, flat panels like this lid on a small box, you can use a bench hook to hold the work.

Illus. 8-26. This bench hook is specifically designed for wood carving. It has stops on two sides and it hooks onto a corner of the bench top. This lets you carve diagonally without causing the panel to slip. Notches in the stops hold square panels at an angle and are also useful for holding irregularly shaped parts.

A carving tray has wedges to secure the work (Illus. 8-27). The tray shown in Illus. 8-28 is large enough for most work, but you can make the tray larger or smaller to suit your needs.

Mount the tray securely to the bench. You can use the front vise dog and a bench dog to secure the tray. You can also use a hold-down. A quick way to mount the carving tray is to attach four dowels to the bottom of the tray. The dowels fit into dog holes in the bench top (Illus. 8-29).

To use the tray, place the work in one corner of the tray and then place wedges between the work and the other sides of the tray. Tighten the wedges by tapping them with a mallet (Illus. 8-30).

VENEERING

When applying veneer to large panels or performing other laminating tasks, you need a way to apply even pressure over the entire face of a panel. One way to accomplish this is to make an accessory veneer press to attach to your workbench. You will need to buy some veneer press screws (Illus. 8-31). You can buy them from a mail-order woodworking-supply company if you can't find them locally.

A small veneer press requires at least four veneer press screws. For larger presses, you will need more screws so that you can space the screws about 8 inches apart.

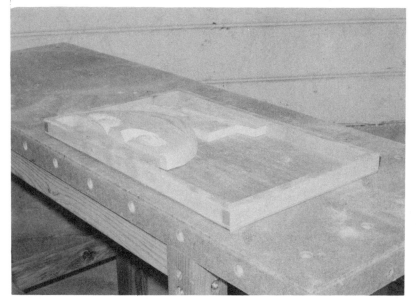

Illus. 8-27. A carving tray has wedges that secure the work. It works well for holding irregularly shaped parts.

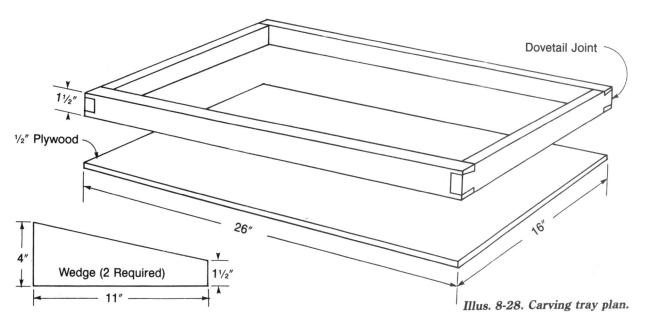

1½"

½" Plywood

Dovetail Joint

26"

16"

4"

Wedge (2 Required)

1½"

11"

Illus. 8-28. Carving tray plan.

Illus. 8-29. Dowels attached to the bottom of the tray fit into dog holes in the bench top.

Illus. 8-30. Tapping the wedges with a mallet secures the work.

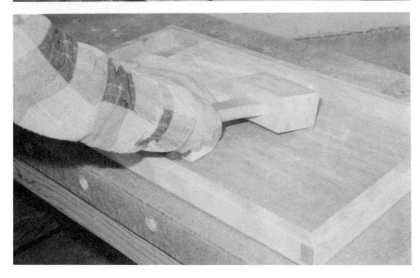

113

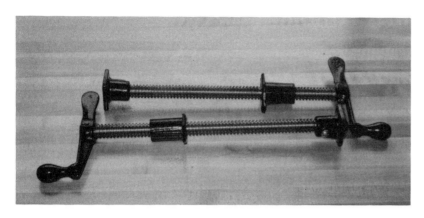

Illus. 8-31. Veneer-press screws like these are available from mail-order woodworking supply companies.

Illus. 8-32 is a plan for a veneer press. The press has threaded rods in its legs with which it is attached to the bench top. Space the legs so that the threaded rods will align with dog holes in the bench top. When the press is in place, install large washers and nuts on the threaded rods.

CHAIRMAKING

For chairmaking you don't need a large bench, but you must be able to hold irregular shapes securely. A small joiner's bench with a twin-screw vise mounted as a front vise works well. Space the vise screws at least 18 inches apart so that a chair seat will fit between them (Illus. 8-33).

To shape the contours of the seat top, clamp the seat with dogs in the vise and bench top. Round dogs are best because they will rotate to the correct angle (Illus. 8-34).

Long, thin strips of wood such as back splats can't be clamped in a tail vise very well, because the vise pressure will bow the middle of the part upwards. Hold-

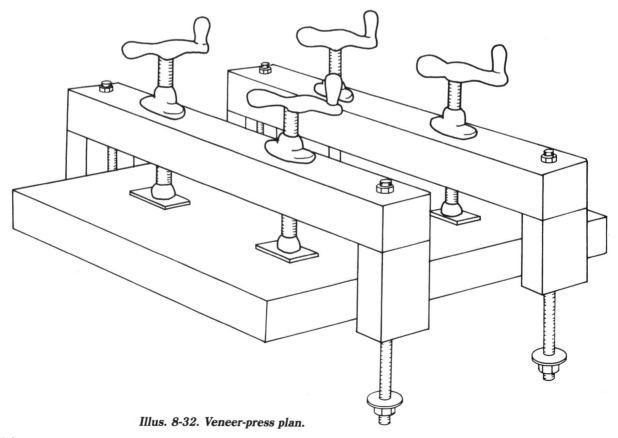

Illus. 8-32. Veneer-press plan.

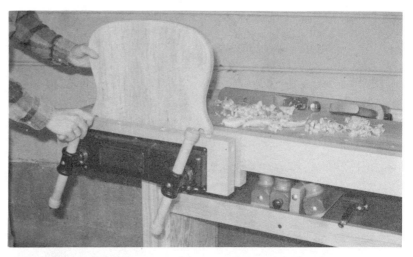

Illus. 8-33. A twin-screw vise mounted in the front vise position works well for chairmaking.

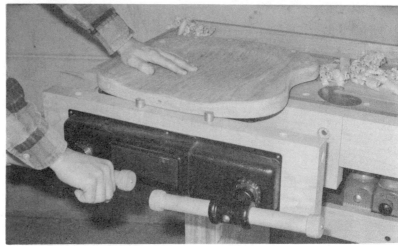

Illus. 8-34. To shape the seat top, clamp the seat with dogs in the vise and bench top.

downs are better for clamping thin wood. For operations such as shaving and planing, place a hold-down at one end and move the tool away from the hold-down. This pulls against the hold-down and keeps the work flat on the bench. If you push into the hold-down, the work will tend to lift and bow.

Illus. 8-35. For shaping parts with a spokeshave or drawknife, clamp the part in a twin-screw vise.

For shaping parts with a spokeshave or a drawknife, place the part in the twin-screw vise (Illus. 8-35). For some work, it is more comfortable to place the work at an angle. The twin-screw vise gives you plenty of vise jaw surface no matter how you angle the work (Illus. 8-36).

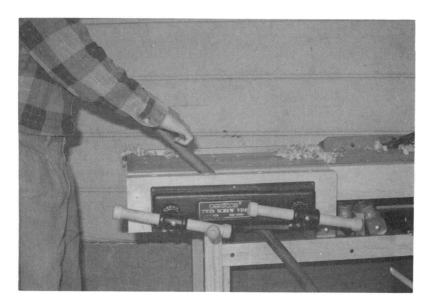

Illus. 8-36. The large, unobstructed jaw area of the twin-screw vise lets you position parts at any angle.

If you do a lot of work with a drawknife, you might want to make a shaving horse (Illus. 8-37). This is the traditional way to hold chair parts as you shape them. You can quickly reposition the work because foot pressure holds the work in place. Illus. 8-38 is a plan for a shaving horse.

Sometimes you need to pull apart the joints in a chair to repair an old chair or correct a mistake as you are building a new chair. You can pull apart chair joints easily with a vise and bench dogs. Reverse the bench dogs from their normal position so that the flat faces of the joints are facing away from each other (Illus. 8-39). Open the vise jaws to pull the joint apart.

Sometimes it is helpful to have vise jaws that extend above the bench top. You can make the vise-jaw extensions shown in Illus. 8-40 and attach them to the twin-screw vise or another type of vise.

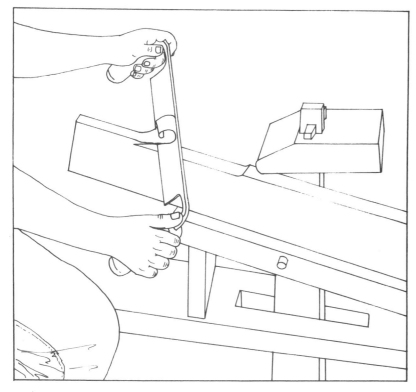

Illus. 8-37. A shaving horse is the traditional way to clamp parts for drawknife and spokeshave work.

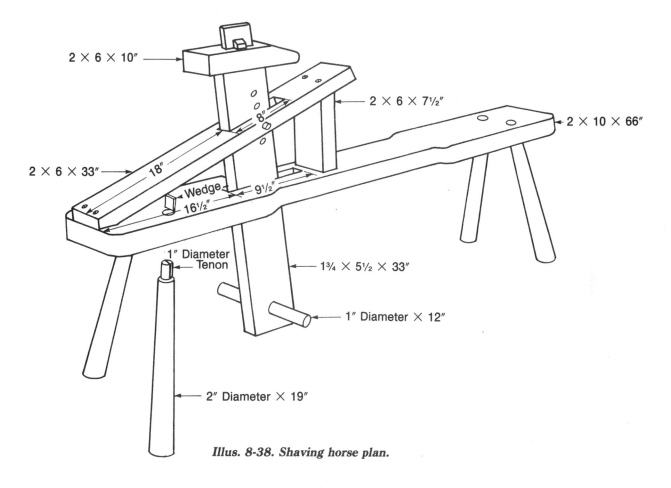

2 × 6 × 10"

2 × 6 × 7½"

2 × 10 × 66"

8"

2 × 6 × 33"

18"

Wedge

9½"

16½"

1" Diameter
Tenon

1¾ × 5½ × 33"

1" Diameter × 12"

2" Diameter × 19"

Illus. 8-38. Shaving horse plan.

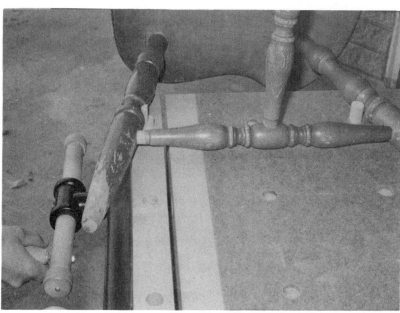

Illus. 8-39. To repair an old chair, you have to pull the joints apart. You can do this with a tail vise. If the legs have flat faces, reverse the dogs so that the flat face rests against the work. In this photo, I've placed the round back of the dog against a turned leg.

Illus. 8-40. These vise-jaw extensions attach to the vise jaws with screws.

GUITAR MAKING AND REPAIRING

A guitar is a fragile project, so you have to be able to secure it to the bench without destroying it. A twin-screw vise works well for holding the body of a guitar edge up (Illus. 8-41). Pad the jaws with cork or carpet to prevent damage to the guitar's finish, and don't tighten the vise too much. You can crush the guitar if you overtighten the vise.

You can use Veritas Wonder Dogs to clamp irregularly shaped parts or to clamp the guitar body flat on the bench top. Pad the face of the Wonder Dogs with cork if you are working on a finished guitar (Illus. 8-42).

A neck on the bench will support the guitar neck and allow you to work on either side of the neck. A saddle on the bench neck will support the guitar neck. Pad the inside of the saddle with felt (Illus. 8-43).

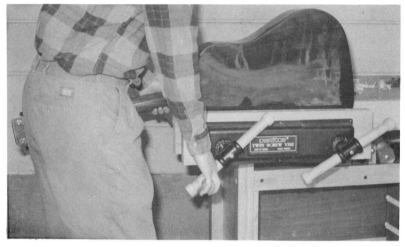

Illus. 8-41. A twin-screw vise will hold a guitar edge-up. Be sure to pad the jaws, and don't tighten the vise too much.

Illus. 8-42. You can use Veritas Wonder Dogs to secure a guitar body flat on the bench top. Pad the face of the Wonder Dogs with cork or felt to prevent damage to the guitar finish. I've pealed back the cork in this photo. Use contact cement to attach the cork.

Illus. 8-43. If you are building a bench specifically for guitar making, you may want to include a neck extension on the bench top. This will support the guitar neck and allow you to work on either side of the neck. The guitar neck rests in a saddle. Pad the inside of the saddle with felt.

CHAPTER NINE

Outfitting Your Bench with Bench Tools

THERE ARE A few tools that are so closely associated with workbenches that they are called bench tools. To get the most out of your workbench, you have to outfit it with a set of bench tools (Illus. 9-1). In this chapter, I describe the common bench tools and explain what they are used for. Table 9-1 is a list of recommended bench tools. This list is not exhaustive. There are many other tools that could be added to the list, but this list represents a fairly complete set of bench tools for the average woodworker. You don't have to buy all of these tools at once.

The tools in the list and described in this chapter are presented to make you aware of what tools are most useful so that you can add to your set of bench tools intelligently as the need arises.

ESSENTIAL BENCH TOOLS

The tools described in this section are essential, so you should probably buy them when you first start woodworking. They are used to lay out your work accurately and make simple joints.

Illus. 9-1. To get the most out of your bench, you have to outfit it with a set of bench tools. The tools laid out on this bench represent some of the common bench tools that are discussed in this chapter.

Recommended Tools For Outfitting a Workbench

Essential Bench Tools

Square
Marking gauge
Block plane
Set of chisels with ¼-, ½-, ¾-, and 1-inch-wide
   blades
Wooden mallet
Backsaw

Additional Tools for Surfacing Lumber

Jack plane
Fore, trying, or jointer plane
Smooth plane

Additional Tools for Joinery

Mortise chisels with ¼- and ⅜-wide blades
Dovetail saw
Rabbet plane
Combination plane

Table 9-1. List of recommended bench tools.

Square

Accurate layout is essential to good woodworking. One of the most important layout tools is the square. There are three types of square that are commonly used. The *framing square* (Illus. 9-2) is a large square usually used by carpenters. It is also useful for laying out large

Illus. 9-2. The framing square is a large square usually used by carpenters. It is also useful for laying out large parts for cabinets.

parts for cabinets. The *try square* (Illus. 9-3) is the traditional cabinetmaker's square. It works well for marking the small parts usually found in cabinetmaking. The *combination square* (Illus. 9-4) is really a machinist tool, but it is so useful that most woodworkers consider it an essential woodworking tool. If you can only buy one square at first, buy a combination square. You can use it to lay out 45-degree mitre joints as well as 90-degree square joints (Illus. 9-5).

Illus. 9-3. The try square is the traditional cabinetmaker's square. It works well for marking the small parts usually found in cabinetmaking.

Illus. 9-4. The combination square is so useful that most woodworkers consider it an essential woodworking tool.

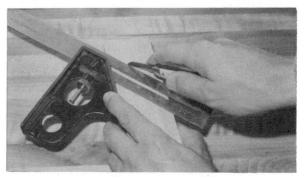

Illus. 9-5. You can use a combination square to lay out 45-degree mitre joints as well as 90-degree square joints.

Marking Gauge

A marking gauge is another layout tool. You can use a marking gauge whenever you need to make a line parallel to one of the surfaces of a board (Illus. 9-6). The standard marking gauge has a pointed steel pin called a spur that scratches a line in the wood. The cutting gauge (Illus. 9-7) is a variation on the marking gauge. It has a sharp blade to cut a line in the wood. The cutting gauge makes a cleaner line across the grain. Scoring the wood fibres with a cutting gauge helps to prevent tear-outs along a cut.

The marking gauge helps to eliminate inaccuracies that can occur when you must make multiple measurements. Set the marking gauge once, and then mark all of the parts that require that measurement. This way, even if you are slightly off from the exact measurement, at least all of the parts will be the same size. Another way you can use the marking gauge to promote accuracy is to take measurements directly from the mating part. When you set the gauge, instead of measuring the mating part and then setting the marking gauge to the measurement, place the marking gauge directly on the mating part and align the spur at the desired location; then set the fence (Illus. 9-8). This eliminates the inaccuracy caused when the measurement you need is between the two marks and you have to make a guess as to where to set the gauge.

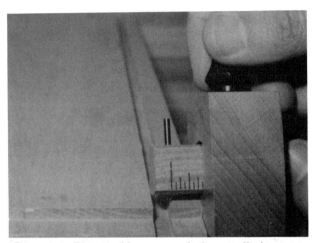

Illus. 9-8. The marking gauge helps to eliminate inaccuracies that can occur when you must make multiple measurements. Set the marking gauge directly from the mating part. Place the marking gauge directly on the mating part and align the spur at the desired location; then set the fence.

Illus. 9-6. The standard marking gauge has a pointed steel pin called a spur, which is used to scratch a line in the wood.

Illus. 9-7. The cutting gauge has a sharp blade that cuts a line in the wood.

Illus. 9-9. The mortise gauge has two spurs, so you can set the gauge to lay out both sides of the mortise or tenon at once.

If you will be making many mortise-and-tenon joints, you should invest in a mortise gauge. This type of marking gauge has two spurs, so you can set the gauge to lay out both sides of the mortise or tenon at once (Illus. 9-9). A mortise gauge makes mortise-and-tenon layout easier and more accurate.

Backsaw

Although most woodworkers today use some type of power saw to do most of their sawing, you still need a good backsaw for cutting joints. A backsaw has a stiffener along the back of its blade, so you can make very accurate cuts with it (Illus. 9-10). You can use a backsaw with the bench hook described in Chapter Five or a mitre box to make accurate square and 45-degree joints.

Illus. 9-10. A backsaw has a stiffenner along the back of its blade, so you can make very accurate cuts with it.

Cutting Tenons with a Backsaw To cut tenons with a backsaw, first lay out the tenon with a mortise gauge or a marking gauge. Next, place the board end up in a vise. To make it easier to follow the layout lines, place the board at about a 45-degree angle in the vise (Illus. 9-11). A shoulder vise is ideal for this operation, because there are no guide rods to get in the way, but you can use any type of vise. Position the saw on the waste side of one of the layout lines on the end of the board. Place your thumb against the side of the saw blade to steady the cut. Start the cut flat across the end. When the cut is about ⅛-inch deep, tilt the saw back until it is parallel with the bench top.

Now, cut straight down following both the layout lines visible on the side and end of the board (Illus. 9-12). Stop when you reach the layout line for the

Illus. 9-11. To make it easier to follow the layout lines, place the board at about a 45-degree angle in the vise when cutting tenons. Place your thumb against the side of the saw blade to steady the cut. Start the cut flat across the end.

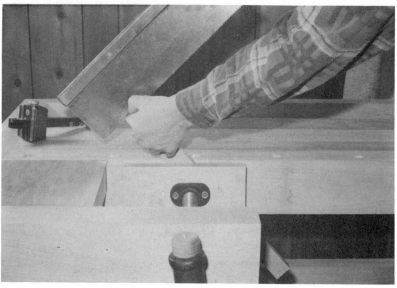

Illus. 9-12. When the cut is about ⅛ inch deep, tilt the saw back until it is parallel with the bench top and cut straight down following both the layout lines, visible on the side and end of the board.

shoulder. Repeat this procedure for the other cheek, and then turn the board over and repeat the procedure again. The final step in cutting the cheeks is to reposition the board straight up in the vise and saw straight down to the shoulder mark (Illus. 9-13).

To cut the shoulders, place the board against a bench hook that has a 90-degree saw-guide-cut or use a mitre box. Place the saw on the waste side of the shoulder layout line and cut straight down (Illus. 9-14).

Block Plane

Many woodworkers buy their lumber already surfaced, so they don't need as many planes as traditional woodworkers who start with a rough board and surface it with hand planes. However, even if you buy all of your wood already surfaced, you will still need a plane to smooth off the saw marks on cut edges and to trim joints. The block plane is well suited to these tasks (Illus. 9-15).

When you cut a mitre joint with a saw, the angle may be slightly off. You can correct this with a block plane. First, assemble the joint and mark the areas that need trimming. Next, place the board in a vise at an angle so that the joint surface is parallel to the bench top. Now, set the block plane for a fine cut and trim the joint to the proper angle. To avoid chipping the edge, start the stroke at the inside corner of the joint (Illus. 9-16). You can also make a guide called a *shooting board* to help you trim a mitre joint (Illus. 9-17). Shooting boards are described in greater detail in my book *Plane Basics.*

Illus. 9-13. The final step in cutting the cheeks is to reposition the board straight up in the vise and saw straight down to the shoulder mark.

Illus. 9-14. To cut the shoulders, place the board against a bench hook and saw on the waste side of the shoulder layout line.

Illus. 9-15. The block plane is good for smoothing off saw marks on cut edges and for trimming joints.

Illus. 9-16. You can trim a mitre joint with a block plane. Place the board in a vise at an angle so that the joint surface is parallel to the bench top. To avoid chipping the edge, start the stroke at the inside corner of the joint.

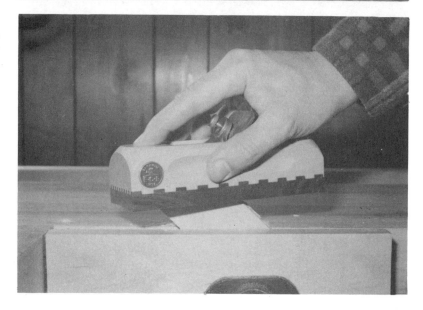

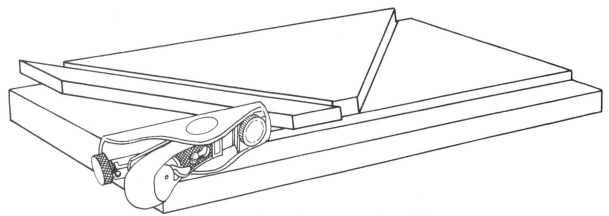

Illus. 9-17. This plan shows how to make a guide called a shooting board. You can use a shooting board to help trim a mitre joint.

Chisels

Chisels are used to chop out the waste between the pins and tails of dovetail joints, to cut short dadoes and grooves, to trim other joints, and to cut gains for hinges and other hardware. Buy a set of quality chisels and they will serve you well for many years. The most commonly used chisels have blades ¼, ½, ¾ and 1 inch wide.

Chisel blades can be classified as either *firmer* or *bevel* blades (Illus. 9-18). Firmer blades have square edges. They are strong and can be used for heavy work. Bevel blades are thinner and have bevelled edges. They are good for getting into tight places and for chopping dovetails, so they are better suited for furniture making.

Illus. 9-18. Chisel blades can be classified as either firmer or bevel blades. Firmer blades (shown on right) have square edges. They are strong and can be used for heavy work. Bevel blades are thinner and have bevelled edges. They are good for getting into tight places.

Chisel blades are also classified by length. The longest blades are about 8 to 10 inches. The chisels themselves are called *patternmaker's chisels.* Patternmaker's chisels are useful if you need to reach deep into an awkward place. *Bench chisels* have blades about 5 to 7 inches long. They are the type most furniture makers use. *Butt chisels* have blades 3 to 4 inches long. They are usually used by carpenters for cutting hinge gains.

Chisels come with several types of handle. The type you choose is really a matter of personal preference because they all perform well. Wood handles are attractive and have a nice feel. Boxwood, beech, and ash are usually used for chisel handles. The handle is either driven over a tang on the end of the blade or fits into a socket formed on the end of the blade. Both types of handle are strong.

Chisels that are made to be struck with a mallet have ferrules to keep the wood from splitting. Chisels used for paring won't have a ferrule at the top. They are meant to be used with hand pressure only. Don't strike a paring chisel with a mallet or you are liable to split the handle.

Plastic handles aren't as traditional looking as wood handles, but they are very strong. High-impact plastic handles don't need ferrules to keep them from splitting, even if you drive them with a mallet. There are several different handle shapes. The choice simply depends on what type of handle feels best in your hand.

Making Chopping Cuts with a Chisel Chisels perform two basic operations: chopping and paring. To use the chisel for chopping, you need one other essential bench tool, a wooden mallet (Illus. 9-19). Striking the chisel with the mallet drives the blade deep into the wood. Chopping is the process of making cuts at ap-

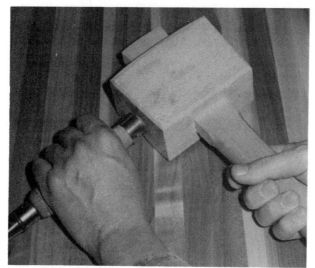

Illus. 9-19. To use the chisel for chopping, you need a wooden mallet. Striking the chisel with the mallet drives the blade deep into the wood.

of the chisel, because the chisel could slip and cut into your hand.

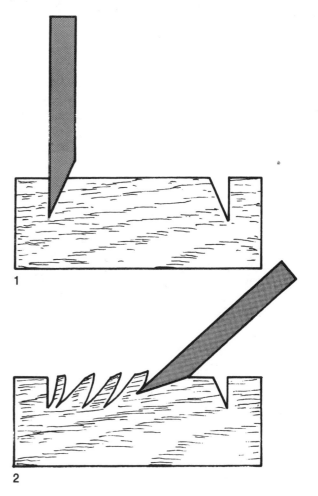

proximately 90 degrees to the face of the board. Chopping can be used to define the edges of a section to be cut out with a chisel. Chopping around the edges cuts the wood fibres so that they won't tear past the line.

Chopping is also useful for removing large amounts of wood and bringing a cut to its approximate depth. Illus. 9-20 shows the process for chopping a recess. The recess could be a hinge gain or part of a joint. Begin by defining the edges by holding the chisel square with the face of the board. Position the cutting edge on the layout line with the bevel facing towards the middle of the recess. Now, drive the chisel into the wood by hitting the handle with the mallet. If the recess is deeper than ¼ inch, just drive the chisel ¼ inch deep for the first series of cuts, and then deepen the recess with another series of cuts.

Next, hold the chisel at about a 60-degree angle with its bevel down. Make a series of parallel cuts across the grain as shown in Illus. 9-20. Then use the chisel to break out the waste wood as also shown in Illus. 9-20 and smooth the bottom of the cut.

Paring with a Chisel Paring is the process of slicing off thin shavings of wood. You can make paring cuts with the bevel up or down. With the bevel down, the handle won't get in the way but you can get a smoother cut with the bevel up. You can use a mallet to drive the chisel when paring, but for very fine cuts you will have more control if you use only pressure from your body to drive the chisel. *Be careful to keep both hands behind the chisel. Don't hold on to the board in front*

Illus. 9-20. The process for chopping a recess. 1. Define the edges by holding the chisel square with the face of the board. Position the cutting edge on the layout line with the bevel facing towards the middle of the recess. Then drive the chisel into the wood by hitting the handle with the mallet. 2. Hold the chisel at about a 60-degree angle with its bevel down. Make a series of parallel cuts across the grain. 3. Use the chisel to break out the waste wood and smooth the bottom of the cut.

127

One important use for paring is removing the waste when cutting joints by hand. First, saw the shoulders of the joint with a backsaw (Illus. 9-21). Then grasp the chisel handle in one hand and extend your index finger along the side. Hold the blade between the thumb and forefinger of your other hand (Illus. 9-22). Put the work in a vise or secure it with hold-downs. You can also press the work against a bench hook or bench stop. Stand back from the work so that you can use your entire body to push the chisel (Illus. 9-23). Don't try to remove all of the waste at once. Paring is a process of removing small shavings in a controlled way. This allows you to make the joint accurately. You can tilt the blade slightly and cut against one shoulder at a time. This leaves a small peak in the middle of the cut. Work down, taking thin shavings from each side of the cut until you reach the layout line, then hold the chisel flat and flatten the bottom of the cut (Illus. 9-24).

A different technique is used to trim tenon shoulders and dovetails. In these situations, you are paring end grain. To pare end grain, hold the chisel with your fingers wrapped around the handle and your thumb on the end of the handle. Guide the blade with the thumb and forefinger of your other hand. Place the work on a bench hook or a scrap board to protect the bench top. The downward pressure from the chisel keeps the work fairly stable even without clamping the board, but for added stability, you can use a hold-down in one of the dog holes (Illus. 9-25).

Lock your arm and bend at the waist as you push against the chisel. With this method you use your body weight instead of your arm muscles to drive the chisel.

Illus. 9-21. One important use of paring is removing the waste when cutting joints by hand. Before you begin to pare with the chisel, saw the shoulders of the joint with a backsaw.

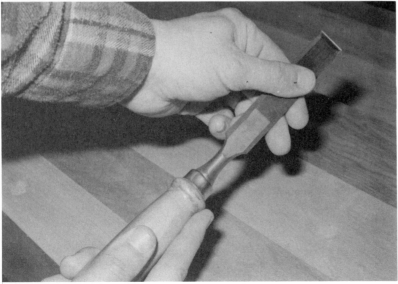

Illus. 9-22. For most paring operations, hold the chisel like this.

Illus. 9-23. You can exert more force with the chisel if you stand back from the work so that you can use your entire body to push the chisel.

Illus. 9-24. Paring is a process of removing small shavings in a controlled way. This allows you to make the joint accurately. You can tilt the blade slightly and cut against one shoulder at a time. This leaves a small peak in the middle of the cut. Work down, taking thin shavings from each side of the cut until you reach the layout line; then hold the chisel flat and flatten the bottom of the cut.

Illus. 9-25. To pare end grain, hold the chisel with your fingers wrapped around the handle and your thumb on the end of the handle. Guide the blade with the thumb and forefinger of your other hand. Place the work on a bench hook or a scrap board to protect the bench top. Lock your arm and bend at the waist as you push against the chisel. With this method, you use your body weight instead of your arm muscles to drive the chisel.

ADDITIONAL TOOLS FOR SURFACING LUMBER

The tools described in this section are more specialized. You only need them if you plan on using hand tools for surfacing lumber.

Bench Planes

If you plan on surfacing your own lumber by hand, you will need some additional planes. Workbenches were originally developed as a way to hold wood for planing, so you will appreciate a well-designed workbench all the more when you use hand planes extensively.

There are many different types of plane available. Bench planes are the most common type. They are used to smooth the face and edges of a board. They are called bench planes because they are usually used in conjunction with a workbench to hold the work. In this section, I describe several important bench planes:

129

the jack plane, the try, fore, and jointer planes, and the smooth plane. For more complete information about choosing, using, and maintaining planes, refer to my book *Plane Basics*.

Planes can be made from wood or metal (Illus. 9-26). Both types have advantages. Metallic planes are readily available, easy to adjust, and durable. Wooden planes are slightly more difficult to find, but a recent resurgence of interest in wooden planes has made it easier to find suppliers that sell wooden planes. Old-time wooden planes were more difficult to adjust than metallic planes, but modern wooden planes incorporate adjustment mechanisms that make them just as easy to adjust as metallic planes. Modern wooden planes also usually have a sole made from one of the very hard woods like hornbeam or lignum vitae. This feature has reduced one of the drawbacks of the older wooden planes. Soles made of softer wood will eventually wear, making the sole uneven and widening the mouth.

Metallic Planes Most metallic planes are designed like the one shown in Illus. 9-27. The body of the plane is called the *stock*. The stock is made from cast metal. The sides of the stock are called *cheeks*, and the bottom of the stock is called the *sole*. The front of the sole is called the *toe*, and the back is called the *heel*. The blade is called a *plane iron*. The *cap iron* attaches to the plane iron. It reinforces the plane iron and acts as a chip breaker. The plane iron sits on the *frog*. The frog holds the plane iron at the proper angle and contains the adjustment mechanism. The opening that the plane iron goes through is called the *throat*, and the visible part of the opening on the sole of the plane is called the *mouth*. The *lever cap* holds the plane iron on the frog.

The *adjustment mechanism* allows you to raise and lower the plane iron and to tilt the plane iron slightly from side to side. The *depth adjusting nut* controls the depth of cut. Turning the nut moves the *Y adjusting lever*. The end of the adjusting lever fits into a slot in the plane iron cap. As the lever moves, it slides the plane iron up or down. The *lateral adjusting lever* tilts the plane iron from side to side. Use the lateral adjusting lever to line up the cutting edge so that it projects uniformly from the mouth. Turn the plane upside-down and sight across the cutting edge as you make this adjustment. Keeping the lateral adjustment correct will help to prevent gouges made by the corners of the plane iron and keep the depth of cut uniform across the width of the plane.

Wooden Planes Wooden planes can have the traditional design shown in Illus. 9-28 or they can incorporate a modern adjustment mechanism as shown in Illus. 9-29. The wooden body of the plane is called the *stock*. The bottom of the stock is called the *sole*. Several types of wood can be used for the stock. Yellow birch, beech, and hornbeam are the most popular types used today.

The front of the stock is called the *fore end*. The opening in the stock where the plane iron is bedded and shavings emerge is called the *throat*. Some planes have a sole made from a harder wood like lignum vitae. The front of the sole is called the *toe*, and the rear of the sole is called the *heel*. The blade opening in the sole is called the *mouth*. Some modern planes have an adjustable mouth opening. This allows you to adjust the size of the opening to suit the size of the shavings.

Over the years there have been many different designs for wooden planes. Today, most planes follow

Illus. 9-26. Planes can be made from wood or metal.

either the German or the English pattern. German planes have a long handle at the fore end called a *horn*. English planes usually have a handle at the rear called a *tote*, and they don't have any handle at the fore end.

Traditional wooden planes have a wooden wedge to secure the plane iron. To adjust the depth of cut on a wedged plane, you must strike the plane with a mallet. To increase the depth of cut, you can either tap the

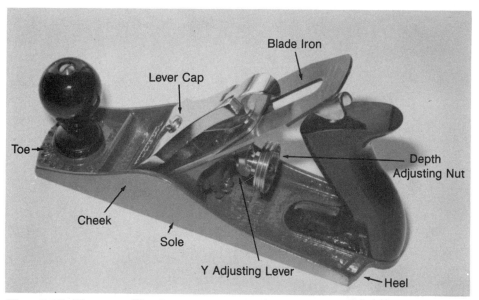

Illus. 9-27. Most metallic planes are designed like the one shown here. The body of the plane is called the stock. *The sides of the stock are called* cheeks, *and the bottom of the stock is called the* sole. *The front of the sole is called the* toe, *and the back is called the* heel. *The blade is called a* plane iron. *The cap iron* attaches *to the plane iron. It reinforces the plane iron and acts as a chip breaker. The plane iron sits on the* frog. *The opening that the plane iron goes through is called the* throat, *and the visible part of the opening on the sole of the plane is called the* mouth. *The lever cap* holds *the plane iron on the frog.*

The adjustment mechanism *allows you to raise and lower the plane iron and to tilt the plane iron slightly from side to side. The depth-adjusting nut* controls the depth of cut. *Turning the nut moves the Y adjusting lever.*

Illus. 9-28. The plane iron on this wooden plane is secured with a wooden wedge. You must use a mallet to adjust it.

Illus. 9-29. This wooden plane has a modern adjustment mechanism that makes it as easy to adjust as a metallic plane.

top of the plane iron with the mallet or tap the toe end of the stock with the mallet (Illus. 9-30). To raise the iron, tap the heel of the plane (Illus. 9-31). Many planes will have a metal or hardwood strike button in the heel so that the mallet blows won't damage the stock. Long planes may have a strike button on the top of their fore end. On these planes, striking the top of the fore end will raise the plane iron. To make lateral adjustments, tap the sides of the plane iron. After adjusting the plane, tap the wedge down to make sure that it is secure (Illus. 9-32).

Illus. 9-31. To raise the iron of a wedged plane, tap the heel of the plane with a mallet.

Illus. 9-30. To adjust the depth of cut on a wedged plane, you must strike the plane with a mallet. To increase the depth of cut, you can either tap the top of the plane iron with the mallet or tap the toe end of the stock with the mallet.

Illus. 9-32. After adjusting the plane, tap the wedge down to make sure that it is secure.

132

Modern wooden planes may have a mechanical adjustment mechanism that eliminates the need to adjust the plane with a mallet. Illus. 9-33 shows the mechanism used in planes made by the E.C.E. company. This mechanism is spring-loaded to eliminate any slack in the adjustment. This is called *zero backlash*. When you make an adjustment, the plane iron begins to move immediately, so you can make very accurate adjustments.

Jack Plane The jack plane is the first plane you use when you are dressing rough lumber. It is used to remove the rough saw marks on the surface of the lumber and to quickly cut the board close to the desired thickness. A jack plane is between 12 and 17 inches long (Illus. 9-34).

For roughing cuts, you can use the jack plane diagonally across the grain of a board. When the plane is used to take deep roughing cuts, its cutting edge can be ground slightly convex. This is called a *cambered* edge. Cutting diagonally across the face of a board with a jack plane with cambered cutting edge will remove a lot of wood quickly, but it will give the board a wavy surface. Once the rough cutting is done, you should switch to another plane to smooth out the waves.

Fore, Trying, and Jointer Planes These long planes can be used to flatten and smooth the surface of a board. They range in length from 18 to 36 inches. The long sole rides on top of the high spots left by the jack plane and slices them off. You can make the first cuts diagonally across the grain and then cut parallel to the grain to smooth the surface. The fore plane is about 18 inches long, the trying plane is about 20 to 24 inches long, and the jointer plane is 22 to 36 inches long. Unless you do a lot of traditional woodwork, you will

Illus. 9-33. This cutaway drawing shows the adjustment mechanism used in planes made by the E.C.E. company. This mechanism is spring-loaded to eliminate any slack in the adjustment. When you make an adjustment, the plane iron begins to move immediately, so you can make very accurate adjustments.

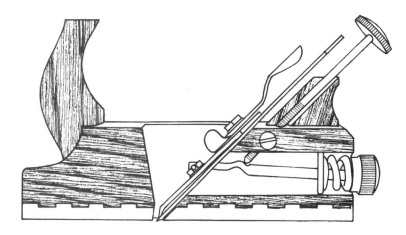

Illus. 9-34. The jack plane is the first plane you use when you are dressing rough lumber. A jack plane is between 12 and 17 inches long.

probably only need one long plane. Any one of these three planes works well for smoothing the surface of a board and jointing edges. I find that a 24-inch long plane suits me best (Illus. 9-35).

Smooth Plane The smooth plane is used to give the board a final smoothing after the other planes have been used. It is also useful for trimming joints and other general planing tasks. A smooth plane is 9 to 10 inches long (Illus. 9-36).

When a board is planed properly with a smooth plane, there is very little need for sanding. The smooth plane leaves a smooth, flat surface. Even if you buy lumber that has already been surfaced, you can benefit by using a smooth plane. Lumber surfaced with power equipment may have a slight wavy pattern on its sur-

face. These waves are called *mill marks*. When the wood is unfinished, the mill marks are almost imperceptible, but when you apply a finish, the mill marks stand out and create an objectionable pattern on the wood surface. Mill marks are often difficult to remove with sandpaper because the sandpaper conforms to the wavy surface. A light planing with a smooth plane will remove mill marks completely.

Planing Technique

In order for you to plane efficiently, the work must be held securely on a stable surface. That is why a good workbench is so important for planing. The tail vise and bench dogs were originally developed as a way to hold a board stationary as the face of the workpiece was planed. Today, they are still one of the best ways

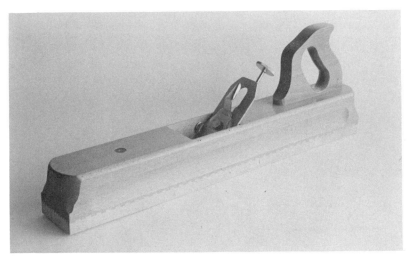

Illus. 9-35. Long planes can be used to flatten and smooth the surface of a board. The fore plane is about 18 inches long, the trying plane is 20 to 24 inches long, and the jointer plane is 22 to 36 inches long. I find that the 24-inch-long trying plane shown here suits me best.

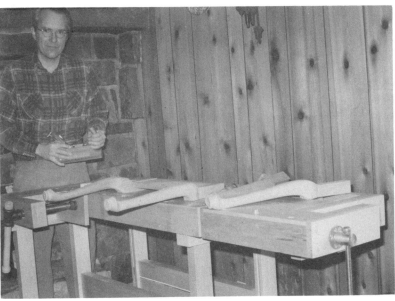

Illus. 9-36. The smooth plane is used to give the board a final smoothing after the other planes have been used. It is also useful for trimming joints and other general planing tasks. A smooth plane is 9 to 10 inches long.

to hold a board securely while you plane it (Illus. 9-37). The front vise can be used to hold a board edge up so you can plane the edge (Illus. 9-38). The vise alone can support short boards, but use a board jack to support the free end of longer boards.

Planing Stance If you get into the proper stance (Illus. 9-39), planing will be more efficient and less tiring. When you plane correctly, your whole body helps move the plane, not just your arms. Stand close to the work with your shoulders directly above the board. Your legs should be spread apart like you were walking. Your forward foot should be almost directly below the board, and your knee should be slightly bent. Your other leg should be almost straight, with your foot slightly behind your body. In this position, you can rock your body while planing to put the force of your body weight into the work. For long cuts, you will have to walk along with the plane. If you start in the proper stance, it is easy to step off into a walk as you plane.

Gripping the Plane Gripping the plane properly gives you the best control of the plane and makes your work more comfortable. Illus. 9-40 shows how to grip a metallic plane. If your hand is too large to fit around the tote, you can extend your index finger alongside the plane iron. To grip a German-style wooden plane, grasp the horn as shown in Illus. 9-41 and wrap your other hand around the heel of the plane. English-style planes don't have a horn. Grasp the fore end of an English-style plane by turning your hand palm forward and placing your fingers on one side of the plane and your thumb on the other (Illus. 9-41). By placing your hand this way, your wrist is free to pivot as the plane moves.

Illus. 9-37. The tail vise and bench dogs are one of the best ways to hold a board securely while you plane it.

Illus. 9-38. The front vise can be used to hold a board edge-up so you can plane the edge.

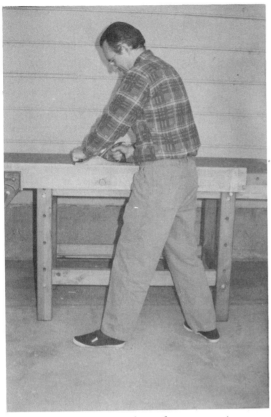

Illus. 9-40. Gripping a metallic plane like this gives you the best control of the plane and makes your work more comfortable. If your hand is too large to fit around the tote, you can extend your index finger alongside the plane iron.

Illus. 9-39. If you get into the proper stance, planing will be more efficient and less tiring. Stand close to the work with your shoulders directly above the board. Your legs should be spread apart as if you were walking. Your forward foot should be almost directly below the board, and your knee should be slightly bent. Your other leg should be almost straight, with your foot slightly behind your body.

Illus. 9-41. To grip a German-style wooden plane, grasp the horn and wrap your other hand around the heel end of the plane.

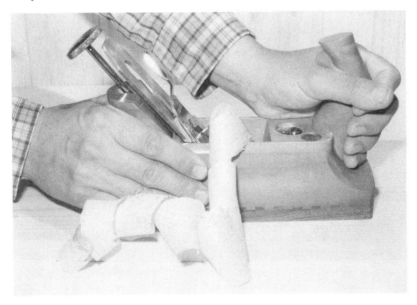

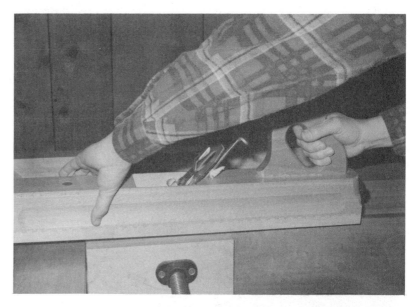

Illus. 9-42. Grasp the fore end of an English-style plane by turning your hand palm forward and placing your fingers on one side of the plane and your thumb on the other. By placing your hand this way, your wrist is free to pivot as the plane moves.

ADDITIONAL TOOLS FOR JOINERY

You can make simple joints with the tools listed above, but if you want to make more advanced joints, you will need some additional tools. Nowadays, many woodworkers use power tools for most joinery operations. For example, you can cut rabbets and dadoes with a router or table saw, make mortises with a router, cut tenons on the table saw, and, with the proper jig, cut dovetails with a router. If you use these power tool methods, then you may not need to buy any of the bench tools listed in this section. However, if you enjoy using hand tools, you may want to use traditional methods for making joints. In that case, the tools described here will be very helpful. If you want detailed instructions for using these tools to make joints, refer to my books *Wood Joiner's Handbook* and *Joinery Basics*.

Mortise Chisels

A mortise chisel has a heavier blade than other chisels, and the blade has a rectangular cross section. The handle is also stronger than other chisel handles so it can endure the heavy pounding that it receives (Illus. 9-43). The rectangular cross section of the blade not only makes the blade stronger, it helps to guide the cut. The sides of the chisel ride against the previously cut parts of the mortise, so if you start out accurately, the mortise chisel will follow the previous cuts to make a straight mortise. Mortise chisels come in several sizes, but for most cabinet work all you really need is a ¼ inch and a ⅜ inch mortise chisel.

Laying Out a Mortise Lay out the mortise with a

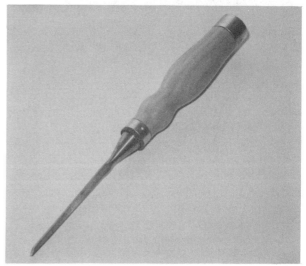

Illus. 9-43. A mortise chisel has a heavier blade than other chisels, and the blade has a rectangular cross section. The handle is also stronger than other chisel handles.

marking gauge. There is a special type of marking gauge called a mortise gauge. The mortise gauge has two spurs, so you can lay out both sides of the mortise at once. This helps to eliminate errors, because the width of the mortise won't be affected by variations in board thickness.

Chopping a Mortise After you have layed out the mortise location, place the board on the workbench. Don't put the board in a front vise, because the repeated blows with a mallet will eventually loosen the vise mounting. Instead, position the mortise on the

137

Illus. 9-44. You can use a tail vise and dogs to hold work as long as you position the board so that the mortise is on the bench top, not on top of the vise.

bench top directly over one of the bench legs. You can use a tail vise and dogs to hold the work, as long as you position the board so that the mortise is on the bench top, not on top of the vise (Illus. 9-44). You can break a tail vise if you pound directly over the vise. A hold-down or a wedge stop can also be used to secure the board (Illus. 9-45).

Start the cut about ⅛ inch in from the end of the mortise layout lines. This will leave you some room to clean up the cut later. The bevel on the chisel should face in towards the middle of the mortise. Hold the chisel away from your body so you can sight along it. The chisel must be straight in both directions.

Hit the handle of the chisel with a mallet. You should be able to drive the chisel about ½ inch deep with each blow from the mallet (Illus. 9-46). Now, move the chisel about ⅜ inch past the first cut and drive it about ½ inch into the wood again. Keep repositioning the chisel about ⅜ inch at a time until you have chopped a series

Illus. 9-45. A wedge stop works well for holding a board while you chop a mortise. It is simple to secure and remove the work with a tap of a mallet.

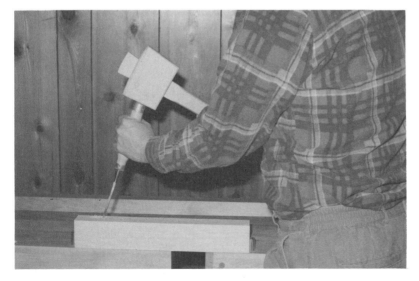

Illus. 9-46. Begin chopping a mortise about ⅛ inch in from the end of the mortise layout lines. Turn the bevel on the chisel towards the middle of the mortise. Hit the handle of the chisel with a mallet. You should be able to drive the chisel about ½ inch deep with each blow from the mallet.

138

of ½ inch deep cuts from one end of the mortise to the other (Illus. 9-47). For the final cut at the opposite end, turn the chisel around so that its bevel faces towards the middle of the mortise. Leave about ⅛ inch between the final cut and the layout line; you will trim to the layout line later.

Illus. 9-47. Chop a series of ½-inch-deep cuts from one end of the mortise to the other.

Now, place the chisel bevel-down at one end of the mortise and drive it to the other end (Illus. 9-48). This will cut off the waste wood and level it out to the depth of the chopping cuts. If the mortise needs to be deeper

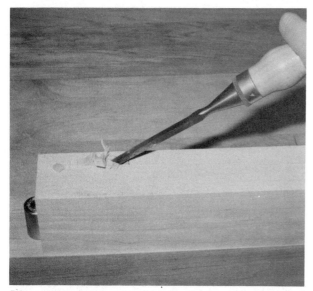

Illus. 9-48. To remove the waste, place the chisel bevel-down at one end of the mortise and drive it to the other end.

than ½ inch, repeat the procedure. Now, the flat sides of the mortise chisel will ride against the sides of the mortise and guide the cut. When the mortise is deep enough, trim its ends to the layout line. Place the cutting edge of the chisel on the layout line and drive the chisel straight down.

Dovetail Saw

Dovetail joints have long been considered the mark of fine craftsmanship. A dovetail joint consists of tapered pins that fit between flared tails (Illus. 9-49). You can cut dovetails with a backsaw, but if you will be cutting many dovetails, you will appreciate the extra control you have with a dovetail saw. A dovetail saw has finer teeth and a thinner blade than a standard backsaw (Illus. 9-50).

Clamp the work end up in the front vise of the workbench and use the dovetail saw to cut the pins and tails (Illus. 9-51). For complete directions for cutting dovetails, refer to my books *Wood Joiner's Handbook* and *Joinery Basics.*

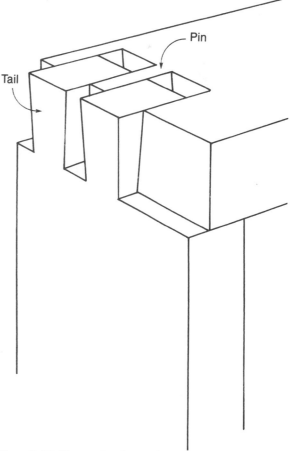

Illus. 9-49. Parts of a dovetail joint.

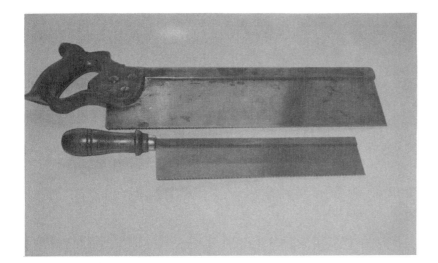

Illus. 9-50. A dovetail saw (bottom) has finer teeth and a thinner blade than a standard backsaw (top).

Illus. 9-51. To cut dovetails, clamp the work end-up to the front vise of the workbench.

Rabbet and Fillister Planes

A *rabbet plane* has a plane iron that extends to the edge of the sole (Illus. 9-52). Rabbet planes are useful for trimming tenon shoulders and for making rabbet joints.

There are many different types of rabbet plane. The simplest is the traditional wooden rabbet plane (Illus. 9-53). This plane can be used to trim shoulders or make rabbet joints. This plane doesn't have any guides to help you make rabbet joints. To guide the plane, wrap your hand around the plane so that your fingers on the sole act as a fence to guide the plane (Illus. 9-54). You can also clamp a board to the face of the work to act as a guide.

A *fillister plane* has a fence and a depth stop to make the job of cutting rabbets easier and more accurate (Illus. 9-55). Clamps and hold-downs can get in the way

Illus. 9-52. A rabbet plane has a plane iron that extends to the edge of the sole.

140

Illus. 9-53. A traditional wooden rabbet plane can be used to trim shoulders or make rabbet joints.

Illus. 9-54. The traditional wooden rabbet plane doesn't have any guides to help you make rabbet joints. To guide the plane, wrap your hand around the plane so that your fingers on the sole act as a fence to guide the plane.

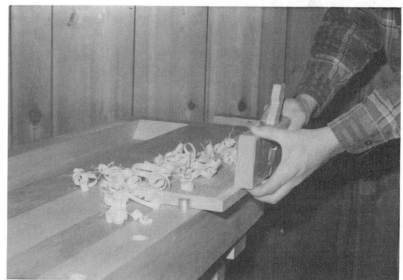

Illus. 9-55. A fillister plane has a fence and a depth stop to make the job of cutting rabbets easier and more accurate.

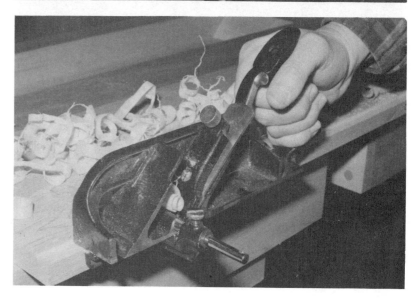

when you use a fillister, so use the tail vise and bench dogs to secure the work when you cut rabbets with a fillister. To prevent the bench top from interfering with the fence make sure that the board overhangs the front edge of the bench top slightly. When you cut a rabbet on a narrow board, you will appreciate having the dog holes as close to the front edge of the bench as possible. The classic cabinetmaker's workbench is designed with this in mind, so you can cut rabbets on a board as narrow as 1¾ inches.

Combination Plane

A combination plane has interchangeable cutters, so you can use it to make rabbets, grooves, dadoes, tongues, and some types of mouldings (Illus. 9-56). For a complete description of the combination plane and how to use it, refer to my book *Plane Basics*.

To cut tongue-and-groove joints on the edge of a board, place the board edge up in the front vise and support the other end with a peg in a board jack (Illus. 9-57).

To cut rabbets and beads, secure the work with the tail vise and bench dogs (Illus. 9-58).

To cut a dado, use hold-downs to secure the work and hold a guide board in position (Illus. 9-59).

Illus. 9-56. A combination plane has interchangeable cutters, so you can use it to make rabbets, grooves, dadoes, tongues, and some types of moulding.

Illus. 9-57. To cut tongue-and-groove joints on the edge of a board, place the board edge-up in the front vise and support the other end with a peg in a board jack.

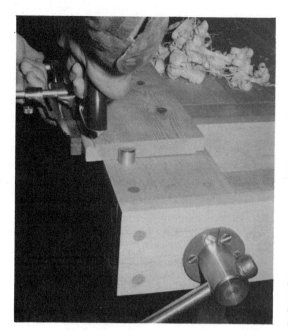

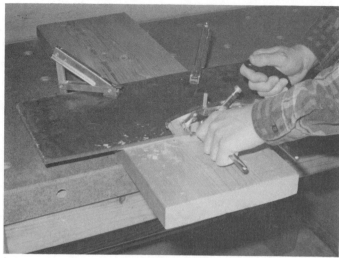

Illus. 9-58 (left). To cut rabbets and beads, secure the work with the tail vise and bench dogs. Illus. 9-59 (above). To cut a dado, use hold-downs to secure the work and hold a guide board in position.

CHAPTER TEN

Outfitting Your Bench with Supplies

T HERE ARE MANY branches of woodworking, ranging from building construction to model making. These specialties include furniture, cabinet, and instrument making. They are referred to under the general term *bench work*.

There are a few supplies that you will frequently use while you work at your bench. To make your bench more useful, you may want to fill a bench caddy or one of the drawers in your bench with some of these supplies. These supplies are described below. Making drawer dividers and a bench caddy in which to store these supplies are described on pages 148–156.

NAILS

There are five types of nail commonly used in woodworking: common, box, finishing, and casing nails and brads (Illus. 10-1).

Common nails have the largest diameter of the five types. They are usually used in carpentry. Their large diameter and large, flat head give them the holding power needed for heavy-duty jobs like house framing. They aren't very useful for bench work. Their large diameter tends to split the thin boards used in bench work.

Like common nails, box nails have a large, flat head, but they are made of thinner gauge wire. You can use them in thinner wood with less chance of splitting. They are sometimes used in cabinetmaking in areas where they won't show. For example, you can use box nails to attach a thin back panel to a cabinet. The large heads on box nails won't pull through the thin panel like a finishing nail head might.

Finishing nails are of even thinner gauge than box nails. They have a small head that can be set below the wood surface. Finishing nails are frequently used to reinforce joints in cabinets and furniture.

Casing nails are designed for attaching door and window casing. They are similar to finishing nails, but they are made from larger, thicker gauge wire and they have a larger, tapering head to give them more holding power. They can be useful when you need added strength, but the larger head leaves a more noticeable hole in the surface.

Brads are the thinnest gauge nails available. They have small heads like finishing nails. When the heads

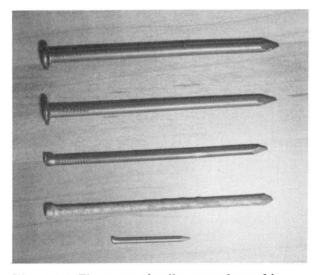

Illus. 10-1. Five types of nail commonly used in woodworking. From top to bottom are common, box, finishing, and casing nails, and a brad.

are set below the surface and the holes filled with putty, they are hardly noticeable.

Nail Sizes

The length of brads is usually designated in inches. Brads are available in lengths that range from ³⁄₈ to 1½ inches. The penny system is used to designate the length of most other nails. The abbreviation for penny is *d*. Standard nail sizes start at 2d, which is about 1 inch long, and end at 60d, which is 6 inches long. For bench work the sizes between 3d and 16d are the most useful.

Nails are usually sold by the pound. Obviously, you will get fewer large nails and more small nails in a pound. Table 10-1 lists the approximate number of nails you can expect to get in a pound, depending on the size and type of nail.

Choosing a Nail

The type and length of nail you choose depends on the appearance of the project, the holding power required, and the thickness of the lumber. To give maximum holding power, a nail should be four times the thickness of the board it is driven through. For example, if you are driving a nail through a ¾-inch-thick board, you will get the maximum holding power with a 10d nail. However, sometimes it isn't possible to use a nail this long. An 8d nail won't have as much holding power, but it is adequate in many situations.

Generally, finishing nails are most often used for bench work, because you don't want an exposed nail head showing on the finished project. In situations where you need added holding power, you can use casing nails. When the heads will be hidden, box nails are a good choice.

Driving Nails

You can usually drive a nail directly into softwoods. To drive a nail into hardwood or thin softwood, you have to drill a pilot hole first. If you don't drill a pilot hole, there is a good chance that the wood will split as you drive the nail. The pilot hole should be slightly smaller than the diameter of the nail.

Nails will hold better in edge grain than in end grain. Unfortunately, when making most joints you have to nail into end grain. You can increase the holding power of the nails by driving them on a slight angle, so that the nails point towards each other. This gives the nails a dovetail effect that makes it more difficult for the boards to be pulled apart (Illus. 10-2).

For most bench work, a 16-ounce hammer works best at driving nails. A lighter hammer will work for small brads, but it will be difficult to drive most nails with a hammer lighter than 16 ounce. Heavier hammers are great for carpentry work, but they don't give you enough control for delicate work like making cabinets and furniture. Hammering against a solid surface makes the job a lot easier (Illus. 10-3). That is where

Common Nail Sizes

Size	Length	Approximate Number Per Pound
2d	1″	845
3d	1¼″	540
4d	1½″	290
5d	1¾″	250
6d	2″	165
7d	2¼″	150
8d	2½″	100
9d	2¾″	90
10d	3″	65
12d	3¼″	60
16d	3½″	45
20d	4″	30
30d	4½″	20
40d	5″	17
50d	5½″	13
60d	6″	10

Finishing Nail Sizes

Size	Length	Approximate Number Per Pound
3d	1¼″	880
4d	1½″	630
6d	2″	290
8d	2½″	195
10d	3″	125

Table 10-1. Nail size chart.

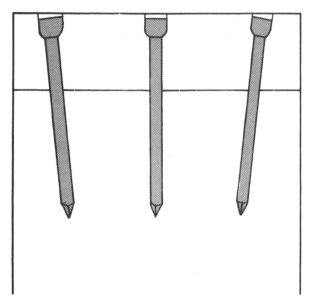

Illus. 10-2. Nails will hold better if you drive them on a slight angle so that the nails point towards each other.

Illus. 10-3. Hammering against a thick bench top makes the job a lot easier. Nails will drive quicker and have less of a chance of bending when you nail on a solid surface.

a thick bench top is very handy. Nails will drive quicker and have less chance of bending when you nail on a solid surface like a thick bench top.

Use a nail set to drive the heads of finishing nails below the surface of the wood (Illus. 10-4). Don't set the nail too deep, or it will lose most of its holding power. If you drive the nail about 1/16 inch below the surface, there will be enough room for the putty.

Illus. 10-4. Use a nail set to drive the heads of finishing nails below the surface of the wood; about 1/16 inch below the surface will provide enough room for the putty.

SCREWS

Screws offer more holding power than nails. The advent of power screwdrivers and new screw types has made the nail almost obsolete in many shops. There are four factors to consider when you choose a screw: size, head type, screwdriver type, and thread type. Each is described below.

Screw Sizes

Screw sizes are designated by a number that indicates the diameter and the length of the screw in inches (Table 10-2). The smallest diameter screw is 0, and the largest commonly available is 24. For bench work, the most useful sizes are 4 through 12. Of those sizes, 6, 8, and 10 are probably used more than any others.

Head Types

Screws have one of three basic types of head: flathead, roundhead, and panhead (Illus. 10-5). Flathead screws are meant to be countersunk flush or below the wood surface. You must countersink the hole before driving the screws (Illus. 10-6 and 10-7). The bugle head screw is a modern variation of the basic flathead screw. It is designed to pull down flush with the surface without being countersunk beforehand. The ordinary bugle head screw will pull down flush in softwoods, but it will remain slightly above the surface in hardwood. Some

Standard Wood Screw Sizes

Screw Size	Approximate Diameter of Screw Shank	Pilot Hole Drill Size
2	3/32″	1/16″
3	3/32″	1/16″
4	7/64″	5/64″
5	1/8″	5/64″
6	9/64″	3/32″
7	5/32″	7/64″
8	11/64″	7/64″
9	3/16″	1/8″
10	13/64″	1/8″
12	7/32″	9/64″
14	15/64″	5/32″
16	17/64″	3/16″
18	19/64″	13/64″

Table 10-2. Screw size chart.

Illus. 10-5. From left to right: flathead, roundhead and panhead screws.

Illus. 10-6. You usually need to drill a pilot hole before driving a screw. Choose a drill bit that is slightly smaller than the root diameter of the screw.

Illus. 10-7. Flathead screws are meant to be countersunk flush or below the wood surface. Use a countersink to cut the recess for the head.

bugle-head screws have sharp nibs below their heads to cut away the wood below the head as you drive the screw. This type will pull flush even in hardwood.

Roundhead screws have a flat bottom, so they don't have to be countersunk. The rounded top extends above the wood surface. Roundhead screws are usually used for attaching hardware. Panhead screws are similar to roundhead screws, but the head is flattened on top so it doesn't extend as much above the surface.

Screwdriver Types

Screws are commonly available with recesses in their heads to fit the following three types of screwdriver: straight blade, Phillips, and square (Illus. 10-8). Screws that take a straight-blade screwdriver are often called *slotted-head screws.* This is the oldest type of screw. Slotted-head screws work fine when you are driving screws by hand, but they are hard to use with a power screwdriver.

The Phillips-head screw has a cross-shaped recess in its head. This type is much easier to drive with a power screwdriver because the bit doesn't jump off the screw head as easily. Phillips-head screws have become increasingly popular as more and more people use power screwdrivers.

The square-head screw has a square recess in its head. Although it is not as widely available as the Phillips-head screw, it is increasingly popular among woodworkers because it offers an even better grip for the screwdriver bit. There are several other types of

147

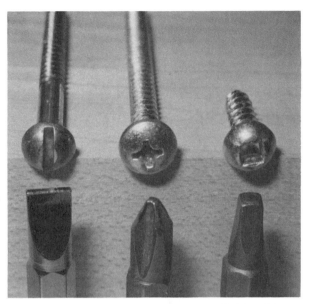

Illus. 10-8. Screws are commonly available with recesses in their heads to fit the three types of screwdriver shown here. From left to right are a straight-blade, Phillips, and square screwdriver.

screwdriver designs, but at present they are mostly used in industrial applications.

Thread Types

The traditional wood screw thread pattern is an old design that was based mostly on the limitations of thread-cutting machinery of a century ago. It has two disadvantages. First, it requires a two-step pilot hole, one size for the threaded portion and a slightly larger size hole for the shank. Second, the threads are fairly shallow, so the screw's holding power is limited.

Sheet-metal screws have deeper and sharper threads, so many woodworkers use them, but a new type of wood screw is quickly becoming very popular. The case-hardened, extruded-thread wood screw has threads that are very deep and sharp (Illus. 10-9). They cut through wood easily and offer a lot of holding power. The shank of the screw is the same size as the root diameter of the threaded portion, so a single-diameter pilot hole can be used. These screws were originally developed for attaching drywall in building construction, so they are often called drywall screws, but new types developed specifically for cabinetmaking are now available. Originally, these screws were only available in a black finish, but now they are available in several types of plated finishes as well. You can drive these screws into softwood and plywood without drilling a pilot hole. For hardwoods, a pilot hole is still recommended.

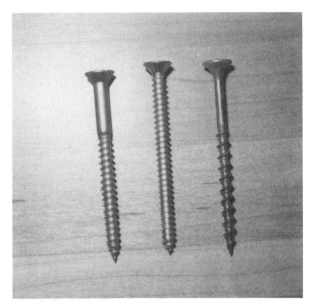

Illus. 10-9. Types of screw thread. From left to right are a traditional wood screw, a sheet-metal screw, and an extruded-thread wood screw.

MAKING DRAWER DIVIDERS

You can organize your supply of nails and screws by adding drawer dividers (Illus. 10-10) to the drawers in the storage base described in Chapter Seven.

The dividers are made from thin plywood or hardboard. Half-lap joints keep the dividers in position. You can cut the half-lap joint with a handsaw (Illus. 10-11) or on a table saw.

Assemble the dividers outside the drawer, and then place the entire assembly in the drawer. Position them so that the lengthwise dividers are on top. This will prevent a potential problem that can occur if the crosswise dividers are on top. If one of the crosswise dividers were to work up, it could jam the drawer closed. By placing the lengthwise dividers on top, none of the crosswise dividers can work its way up. A thin strip of wood screwed to the front and back of the drawer holds the lengthwise dividers in position (Illus. 10-12).

BUILDING A BENCH CADDY

A bench caddy is a small tray with dividers. It is a handy bench accessory for keeping nails and screws organized. If you took junior high school wood shop around the time I did, you may recognize this project. It was commonly required as the first project in woodworking class. Industrial arts programs with new technology have replaced woodworking in many junior high school districts, so I have included plans here for those

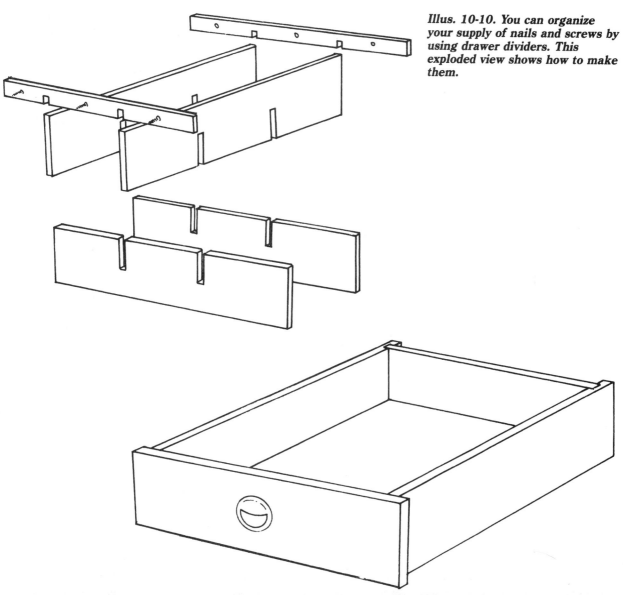

Illus. 10-10. You can organize your supply of nails and screws by using drawer dividers. This exploded view shows how to make them.

Illus. 10-11. The dividers are made from thin plywood or hardboard. Half-lap joints keep the dividers in position. You can cut the half-lap joint with a handsaw.

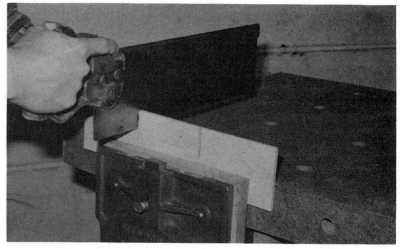

Illus. 10-12. A thin strip of wood attached to the front and back of the drawer holds the lengthwise dividers in position. Cut a notch for each of the dividers. Attach the strip to the drawer with small screws.

of you who didn't get to build one in junior high school.

The bench caddy shown in Illus. 10-13 is the first woodworking project I ever built. I made it over 30 years ago in seventh-grade woodworking class. This is a good project to make when you are learning how to use hand tools because it incorporates practically all of the basics of woodworking with hand tools in a small project. If you're new to woodworking with hand tools, this project will give you some useful practice with your new workbench and bench tools.

Start by cutting some ¾-inch-thick lumber to the widths shown in Illus. 10-14. Although you can use a ripsaw to cut the wood to width, all but the most traditional hand-tool woodworkers probably will cut the wood to width with a table saw. Planing will be easier if you cut one strip of wood long enough for both sides and both ends, then cut the parts to length after planing.

Next, you have to plane the wood to the required ½-inch thickness. Set a marking gauge to ½ inch and scratch a line around the edges of the boards. Clamp the board down on the bench top using the tail vise and bench dogs. On boards this small, you won't need a long plane. You can do all of the planing with a jack plane (Illus. 10-15).

Before you cut the parts to length, cut the rabbet for the bottom of the bench plane with a rabbet plane. Position the board so that the edge slightly overhangs the bench top and clamp it with the tail vise and bench dogs (Illus. 10-16).

Now, cut the parts to length. Use a square to lay out the cuts. Then use a backsaw and a bench hook to cut the parts to length.

Next, cut the rabbets and dadoes. Lay out the locations with a square. Make the shoulder cuts with a backsaw. Use a bench hook to saw against (Illus.

Illus. 10-13. If you took junior high school wood shop around the time I did, you may recognize this project. I built this bench caddy over 30 years ago in seventh-grade woodworking class. This is a good project to make when you are first starting with hand tools, because it incorporates practically all of the basics of woodworking with hand tools in a small project. If you're new to woodworking with hand tools, this project will give you some useful practice with your new workbench and bench tools.

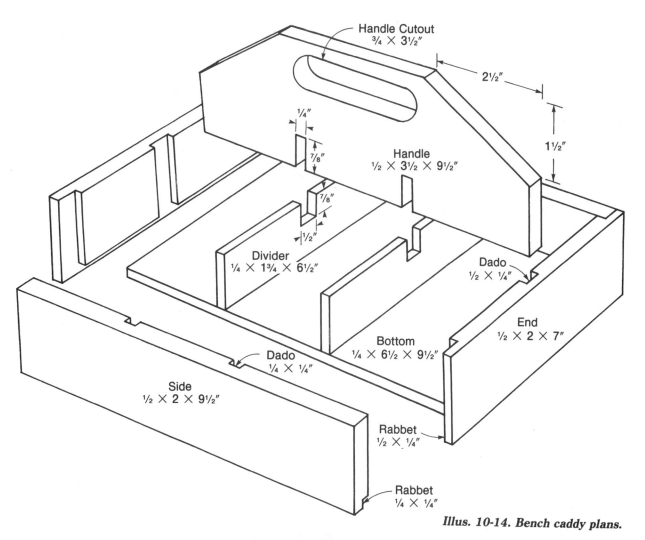

Handle Cutout
¾ × 3½"

2½"

1½"

¼"

⅞"

Handle
½ × 3½ × 9½"

⅞"

½"

Divider
¼ × 1¾ × 6½"

Dado
½ × ¼"

Bottom
¼ × 6½ × 9½"

End
½ × 2 × 7"

Dado
¼ × ¼"

Side
½ × 2 × 9½"

Rabbet
½ × ¼"

Rabbet
¼ × ¼"

Illus. 10-14. Bench caddy plans.

Illus. 10-15. Plane the wood to the required ½-inch thickness. To do this, set a marking gauge to ½ inch and gauge a line around the edges of the boards. Clamp the board down on the bench top using the tail vise and bench dogs. On boards this small, you won't need a long plane. You can do all of the planing with a jack plane.

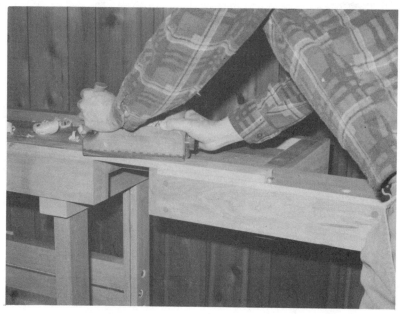

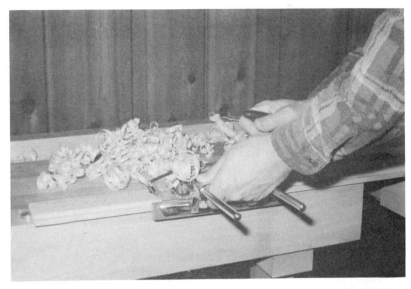

Illus. 10-16. Before you cut the parts to length, cut the rabbet for the bottom of the bench caddy with a rabbet plane. Position the board so that its edge slightly overhangs the bench top, and clamp it with the tail vise and bench dogs.

Illus. 10-17. Use a backsaw and a bench hook to cut the shoulders of the rabbets and dadoes.

Illus. 10-18. Place the parts end-up in the front vise and make the cheek cuts on the rabbets with a backsaw.

10-17). Place the parts end up in the front vise and make the cheek cuts on the rabbets with a backsaw (Illus. 10-18). Then use a chisel to pare away the waste between the shoulder cuts of the dadoes (Illus. 10-19).

Now, turn your attention to the handle. Drill ¾-inch holes at each end of the handhold cutout, and then cut

Illus. 10-19. Use a chisel to pare away the waste between the shoulder cuts of the dadoes.

Illus. 10-20. Drill ¾-inch holes at each end of the handhold cutout, and then cut between the holes with a coping saw.

Illus. 10-21. Saw the top corners off at an angle.

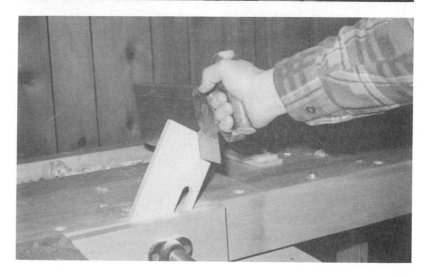

Illus. 10-22. Smooth the rough edge left by the saw with a block plane.

between the holes with a coping saw (Illus. 10-20). Next, saw the top corners off at an angle (Illus. 10-21). You can smooth the rough edge left by the saw with a block plane (Illus. 10-22). Finally, cut the half-lap joints for the dividers. Use the backsaw to make the two side cuts, and then chop out the waste with a chisel (Illus. 10-23).

The two dividers are made from ¼-inch-thick plywood. Cut them to length with a backsaw and bench hook, and then put them in the front vise and cut the half-lap joints. You can cut the joints in both boards at once by clamping them together in the vise (Illus. 10-24).

Now, you are ready to assemble the parts. Start by attaching the ends to the handle. Apply some glue in the dado and assemble the joint. To make assembly easier, clamp the handle in the front vise while you attach the first end. Drive two 4d finishing nails through the end into the handle (Illus. 10-25).

After attaching the second end, insert the dividers in the half-lap joints in the handle. Next, apply glue to the rabbet joints and attach the sides. Use two 4d finishing nails to secure each joint (Illus. 10-26).

To make it easier to nail the bottom in place, turn the caddy upside down and clamp the handle in the front vise. Let the side of the caddy rest on the bench

Illus. 10-23. Use a chisel to chop out the waste from the half-lap joints.

154

Illus. 10-24. You can cut the joints in both of the plywood dividers at once by clamping them together in the vise.

Illus. 10-25. Start assembling the caddy by attaching the ends to the handle. Apply some glue in the dado and assemble the joint. You can clamp the handle in the front vise while you attach the first end to make assembly easier. Drive two 4d finishing nails through the end into the handle.

Illus. 10-26. After attaching the second end, insert the dividers in the half-lap joints in the handle. Next, apply glue to the rabbet joints and attach the sides. Use two 4d finishing nails to secure each joint.

155

top. Apply glue to the rabbet and insert the ¼-inch plywood bottom. Use ¾-inch brads to secure the bottom in the rabbet. Drive the brads on the side that is resting on the bench top. Then loosen the vise and turn the caddy around before you drive the brads in the other edge (Illus. 10-27).

Illus. 10-27. To make it easier to nail the bottom in place, turn the caddy upside down and clamp the handle in the front vise. Let the side of the caddy rest on the bench top. Drive the brads on the side that is resting on the bench top. Then loosen the vise and turn the caddy around before you drive the brads in the other edge.

METRIC EQUIVALENCY CHART

mm—millimetres   **cm—centimetres**

INCHES TO MILLIMETRES AND CENTIMETRES

inches	mm	cm	inches	cm	inches	cm
⅛	3	0.3	9	22.9	30	76.2
¼	6	0.6	10	25.4	31	78.7
⅜	10	1.0	11	27.9	32	81.3
½	13	1.3	12	30.5	33	83.8
⅝	16	1.6	13	33.0	34	86.4
¾	19	1.9	14	35.6	35	88.9
⅞	22	2.2	15	38.1	36	91.4
1	25	2.5	16	40.6	37	94.0
1¼	32	3.2	17	43.2	38	96.5
1½	38	3.8	18	45.7	39	99.1
1¾	44	4.4	19	48.3	40	101.6
2	51	5.1	20	50.8	41	104.1
2½	64	6.4	21	53.3	42	106.7
3	76	7.6	22	55.9	43	109.2
3½	89	8.9	23	58.4	44	111.8
4	102	10.2	24	61.0	45	114.3
4½	114	11.4	25	63.5	46	116.8
5	127	12.7	26	66.0	47	119.4
6	152	15.2	27	68.6	48	121.9
7	178	17.8	28	71.1	49	124.5
8	203	20.3	29	73.7	50	127.0

Index